NATURAL AND INDUCED
CELL-MEDIATED CYTOTOXICITY

Perspectives in Immunology
A Series of Publications Based on Symposia

Maurice Landy and Werner Braun (Eds.)
IMMUNOLOGICAL TOLERANCE
A Reassessment of Mechanisms of the Immune Response
1969

H. Sherwood Lawrence and Maurice Landy (Eds.)
MEDIATORS OF CELLULAR IMMUNITY
1969

Richard T. Smith and Maurice Landy (Eds.)
IMMUNE SURVEILLANCE
1970

Jonathan W. Uhr and Maurice Landy (Eds.)
IMMUNOLOGIC INTERVENTION
1971

Hugh O. McDevitt and Maurice Landy (Eds.)
GENETIC CONTROL OF IMMUNE RESPONSIVENESS
Relationship to Disease Susceptibility
1972

Richard T. Smith and Maurice Landy (Eds.)
IMMUNOBIOLOGY OF THE TUMOR—HOST
　RELATIONSHIP
1974

Gustavo Cudkowicz, Maurice Landy, and Gene M. Shearer
　(Eds.)
NATURAL RESISTANCE SYSTEMS AGAINST FOREIGN
　CELLS, TUMORS, AND MICROBES
1978

Gert Riethmüller, Peter Wernet, and Gustavo Cudkowicz
　(Eds.)
NATURAL AND INDUCED CELL-MEDIATED
　CYTOTOXICITY
Effector and Regulatory Mechanisms
1979

NATURAL AND INDUCED CELL-MEDIATED CYTOTOXICITY
Effector and Regulatory Mechanisms

edited by

GERT RIETHMÜLLER *University of Munich*
Munich, Germany

PETER WERNET *University of Tübingen*
Tübingen, Germany

GUSTAVO CUDKOWICZ *State University of New York*
Buffalo, New York

Proceedings of the Erwin Riesch Symposium
Organized on the Occasion of the Fifth Centennial
of the University of Tübingen
at Weitenburg bei Tübingen, Germany
October 20–23, 1977

ACADEMIC PRESS   1979

A Subsidiary of Harcourt Brace Jovanovich, Publishers
New York   London   Toronto   Sydney   San Francisco

ACADEMIC PRESS, INC.
111 Fifth Avenue, New York, New York 10003

United Kingdom Edition published by
ACADEMIC PRESS, INC. (LONDON) LTD.
24/28 Oval Road, London NW1   7DX

Library of Congress Cataloging in Publication Data

Erwin Riesch Symposium, Tübingen, 1976.
    Natural and induced cell-mediated cytotoxicity.

    1. Cellular immunity—Congresses.   2. Immunocompet-
ent cells—Congresses.   I. Riethmüller, Gert.
II. Wernet, Peter.   III. Cudkowicz, Gustavo.   IV. Title.
[DNLM: 1. Cytotoxicity, Immunologic—Congresses.
2. Immunity, Cellular—Congresses.   QW568 N285]
QR185.5.E78 1976   599'.02'9   79-14162
ISBN 0-12-584650-9

PRINTED IN THE UNITED STATES OF AMERICA

79 80 81 82   9 8 7 6 5 4 3 2 1

CONTENTS

INDUCED CELL-MEDIATED CYTOTOXICITY:
TARGETS, MECHANISMS, AND CONCEPTS

INDUCED CELL-MEDIATED CYTOTOXICITY:
REACTIVITY AGAINST MODIFIED SELF IN MAN

MACROPHAGES AS REGULATOR, ACCESSORY, AND EFFECTOR CELLS IN CYTOTOXICITY

CONCLUSIONS AND OVERVIEW

CONFEREES

Guy D. Bonnard   National Cancer Institute, Bethesda, Maryland

W. Hallowell Churchill   Harvard Medical School, Boston, Massachussetts

Gustavo Cudkowicz   State University of New York, Buffalo, New York

Peter Dukor   Ciba–Geigy AG, Basel, Switzerland

Kirsten Fischer-Lindahl   University of Cologne, Cologne, Federal Republic of Germany

Hans D. Flad   University of Ulm, Ulm, Federal Republic of Germany

Bernhard Fleischer   Justus von Liebig University, Giessen, Federal Republic of Germany

Roland Gisler   Ciba–Geigy AG, Basel, Switzerland

Martin Hadam   University of Munich, Munich, Federal Republic of Germany

Otto Haller   University of Zürich, Zürich, Switzerland

Stephen Haskill   University of South Carolina, Charleston, South Carolina

Dominique Juy   Institut Pasteur, Paris, France

Rolf Kiessling   Karolinska Institute, Stockholm, Sweden

Holger Kirchner   German Center for Cancer Research, Heidelberg, Federal Republic of Germany

Eva Klein   Karolinska Institute, Stockholm, Sweden

Ulrich Koszinowski   University College London, London, England

Marie-Luise Lohmann-Matthes   Max–Planck Institute of Immunobiology, Freiburg, Federal Republic of Germany

Arpi Matossian-Rogers   London Hospital Medical College, London, England

Andrew McMichael   University of Oxford, Oxford, England

N. Avrion Mitchison   University College London, London, England

Malcolm A. S. Moore   Memorial Sloan–Kettering Cancer Center, New York

Paul G. Munder   Max–Planck Institute of Immunobiology, Freiburg, Federal Republic of Germany

Roderick T. D. Oliver   National Institute for Medical Research, Mill Hill, London, England

Gerd R. Pape   University of Stockholm, Stockholm, Sweden

Peter Perlmann   University of Stockholm, Stockholm, Sweden

Hans H. Peter   Medical College, Hannover, Federal Republic of Germany

Hugh F. Pross   Ontario Cancer Foundation, Kingston, Ontario, Canada

Ernst P. Rieber   University of Munich, Munich, Federal Republic of Germany

Gert Riethmüller   University of Munich, Munich, Federal Republic of Germany

E. J. Ruitenberg   National Institute of Public Health, Bilthoven, The Netherlands

Johannes G. Saal   Robert Bosch Hospital, Stuttgart, Federal Republic of Germany

Volker Schirrmacher   German Center for Cancer Research, Heidelberg, Federal Republic of Germany

Martin T. Scott   Wellcome Research Laboratories, Beckenham, England

Fritz R. Seiler   Research Laboratories von Behringwerke, Marburg, Federal Republic of Germany

Gene M. Shearer   National Cancer Institute, Bethesda, Maryland

Mitsuo Takasugi   University of California, Los Angeles, California

Giorgio Trinchieri   Swiss Institute for Experimental Cancer Research, Lausanne, Switzerland

Marita Troye   University of Stockholm, Stockholm, Sweden

Hermann Wagner   Johann Gutenberg University, Mainz, Federal Republic of Germany

Dorothee Wernet   University of Tübingen, Tübingen, Federal Republic of Germany

Peter Wernet   University of Tübingen, Tübingen, Federal Republic of Germany

PREFACE

The past five years have seen the emergence and rapid development of information on spontaneously occurring (natural) cytotoxic cells active against tumor targets of various histologic types. While seeming to be a startling new phase of tumor immunology, scanning of the older literature reveals a considerable number of individual observations and reports whose significance was not then appreciated, but are now seen as early antecedents of what are presently termed NK (natural killer) and K (killer) cells. Other examples of natural cytotoxicity against nonneoplastic targets are also to be found, notably the resistance of irradiated animals to the engraftment of hemopoietic stem cells as well as the genetically controlled natural resistance of experimental animals manifested against a variety of microbial pathogens.

Toward the end of 1976, a conference* on Natural Resistance Systems was organized in which for the first time these seemingly disparate noninduced phenomena were critically evaluated as to cellular mechanism and regulation. It was appreciated that natural host responses reflected a good deal more than "background" activities attributable to environmental stimuli. Attention was directed to the attributes shared by these varied systems and the possibility was recognized that they might well be facets of a common basic mechanism. Since then, additional experimentation led to considerable progress in delineating natural cytotoxic cells selective for transformed targets of lymphoid origin, not only in rodents but in human systems as well. Subsequently, we felt that these concurrent developments warranted special emphasis in the form of an Erwin Riesch Symposium celebrating the 500th anniversary of the University of Tübingen. In planning the conference, it was sought to bring together the latest developments on natural and induced cytotoxicity and on

---

*"Natural resistance systems against foreign cells, tumors, and microbes." *In* "Perspectives in Immunology," Volume VII (G. Cudkowicz, M. Landy, and G. M. Shearer, eds.). Academic Press, New York, 1978.

nonspecific immunopotentiation, since mechanisms operative in the latter might well be relevant to a better understanding of natural resistance. Furthermore, it was sought to review for the first time these natural phenomena in human systems and to assess the expanding technology serving this area. The conference itself affirmed the considerable progress made in understanding natural cytotoxic cells, but the other expectation—that work on immunostimulants would materially increase knowledge of natural resistance mechanisms and, indeed, how immunostimulants themselves function—was not realized. Investigations in the field of immunopotentiation are still largely based on concepts and models of induced immunity. It is primarily for this reason that the extensive body of research on immunostimulants has failed so far to contribute to the current understanding of natural immunity. Consequently, in these edited conference proceedings the presentations based on immunostimulation, lacking direct reference to noninduced cytotoxicity, have been omitted.

Despite the considerable effort and achievements in work on mechanisms of natural resistance, the emphasis contemporary immunology puts on induced immunity mediated by cytotoxic T lymphocytes far overshadows work with noninduced effector cells, such as is reported here. In the foreground of present developments, there continues to be a great preoccupation with natural cytotoxicity against tumor targets; this in turn results in excessive rationalization for these cells performing a surveillance role. The limits of our understanding are severely constrained by our continuing efforts to interpret the significance of these effectors solely in terms of the targets used in experimental models. It seems to us more objective to seek the truly basic function of such a pervasive system of natural reactivities and autoreactivity. The conference itself and a number of recent reports have given what we regard as clues to the basic character of natural killer function, i.e., modulation or regulation of proliferative activity and differentiation in cell renewal systems. A picture is emerging of the broad classes of NK cells selective in their recognition of various targets. The basis for such recognition as well as the nature of the relevant target determinants are clearly separate and distinct from immune T and B lymphocyte recognition, the touchstone of immunology. At the conference, this issue was identified as *the* central unresolved element presently precluding a real understanding of natural host responses.

G. RIETHMÜLLER
P. WERNET
G. CUDKOWICZ

ACKNOWLEDGMENTS

The conference organizers gratefully acknowledge financial support from the Erwin Riesch Foundation, Lorch, Germany, and the Jubilee Fund of the University of Tübingen, Germany. Supplemental contributions from the following foundations and companies are also acknowledged: Paul Martini Foundation; Bayer AG; Behringwerke AG; Deutsche Wellcome GmbH; and Searle GmbH.

The conference organizers especially thank Dr. Peter Dukor and Dr. Peter Perlmann for help in formulating the conference program. The participation and advice of Dr. Maurice Landy, Schweizerisches Forschungsinstitut, Davos, Switzerland, in the editing of the conference proceedings was invaluable.

CHARACTERISTICS OF NATURAL CYTOTOXIC CELLS IN MICE

Rolf Kiessling

Several groups of investigators have shown that tumor cells can be destroyed by effector cells from normal donors who, as far as could be determined, had neither been immunized nor sensitized. These effector cells have been designated NK, i.e., natural killer cells, in analogy with the phenomenon of natural antibodies. NK-cell phenomena have been studied in mice, rats, and man. Almost all studies have utilized the short term ^{51}Cr release assay, and most of the experiments to be considered have been performed with this technique. In the case of rats and man, long-term assays, such as the microcytotoxic assays, have been employed.

SPECIFICITY OF MOUSE NK CELLS

It has been concluded by several groups of investigators that mouse NK cells show a reproducible pattern with respect to target cell sensitivity (1-4). T lymphomas constitute the tumor cell type most susceptible to NK lysis, but other tumors of nonhematopoietic origin also express significant lytic sensitivity (1, 2). Susceptibility to lysis by NK cells depends on expression of relevant target structures on the various cell lines as verified in competition assays. In these assays the capacity of various "cold" competitor cells to inhibit isotope release from labeled target cells was assessed. A good correlation between susceptibility to direct lysis and the ability to inhibit lysis was generally found (1, 2). NK-cell-mediated lysis was found to function across H-2 (1) or species barriers (5).

Target cells cultured *in vitro* are consistently more sensitive to NK lysis than

1

the respective *in vivo* lines (1, 2). Following *in vitro* explantation of a mouse lymphoma, this increase in lysis sensitivity did not occur until after three weeks of *in vitro* culture.

A variety of cell surface components have been considered as the target site for the cytolytic activity of NK cells, but no compelling evidence is available to support any of the hypothetical explanations. The most widely supported notion suggested that NK activity is directed against endogenous C-type viral antigens on the target surface (2). However, Becker and Klein (6) found no correlation between the NK susceptibility of Moloney lymphoma sublines that differed markedly in Moloney virus-determined antigen expression. It is also noteworthy that mouse NK cells can kill certain human targets, T-cell lymphomas, in particular (5). Furthermore, human cell lines that were deliberately superinfected with mouse xenotropic C-type virus by passage in nude mice and subsequently tested for NK sensitivity showed no increase or altered sensitivity to mouse NK cells. Thus, the target specificity attacked by the NK cell remains unknown. It is of interest that recent evidence suggests that NK cells can play a role in resistance not only to certain tumor lines, but also to normal bone marrow grafts (7, 8). Accordingly, NK cells could also play a role in the control of hemopoietic differentiation. In that event they probably recognize a cell type associated with other than viral specificities.

IN VITRO CHARACTERISTICS OF MOUSE NK CELLS

It is now well established that in the mouse the NK cell is not a mature T cell (9, 10) as it lacks detectable amounts of theta-antigen and inasmuch as nude mice exhibit high activity. Furthermore, mouse NK cells lack surface immunoglobulin as well as C3 receptors (9, 10). Despite earlier negative findings (10), Herberman *et al.*, (11) have concluded that NK cells have low but detectable amount of Fc receptors. There is general agreement that the NK cell is poorly adherent and nonphagocytic (3, 4, 9, 10). Accordingly, NK cells in mice are neither mature B nor T cells. Reports describing the rat NK cell seem to be in concurrence with those from mouse systems (12, 13). Here too, the killer cell seems to be of non-T origin as measured by a heterologous anti-T serum. Also, rat NK cells lack C3 and Fc receptors, and are poorly adherent and nonphagocytic.

Even this brief summary suffices to conclude that mouse and rat NK cells constitute a cell type clearly distinguishable from previously defined cytotoxic cells.

NONGENETIC FACTORS INFLUENCING NK ACTIVITY

NK activity in the mouse can be influenced by a number of factors as listed in Table 1.

First to be mentioned are some nongenetic methods used to manipulate NK

TABLE 1. Influence of Nongenetic and Genetic Factors

Age: Newborn low, onset 3-4 weeks, peak 6-8 weeks, decline 3-6 months
    Influence of nongenetic factors:
        Suppressed in tumor bearing mice
        Augmentation by injection of tumor cells, bacterial adjuvants and
            acute virus infection

    Influence of genetic factors:
        Strain dependent
        High-reactivity dominant
        H-2 linked influence

activity. The pronounced influence on NK activity of the age of the animal is noteworthy. Fetal liver and spleen cells from new-born mice show little or no activity. In contrast to immune T cells, which in mice mature during the first week of life, the onset of NK activity does not occur until three weeks of age, and peak activity is seen in 6-8 week old mice. Thereafter there was a marked decline of activity, 6-12 months old mice showing very low activity (1, 2). In rats a roughly similar situation has been observed (12, 13).

There are several lines of evidence that NK cells are of major importance for rejection of subcutaneously injected tumor cells (14, 15). NK cells can probably migrate from the spleen or peripheral blood, organs known to contain the highest NK activity, into the local site of tumor growth. Alternatively, they could be recruited from cells already present in various organs. It will be important to ascertain the influence of tumor induction and of immunization with tumor cells on NK activity in peripheral lymphoid organs. Becker and Klein (6) have studied the NK activity in three separate models of tumor-bearing animals, and in all three systems a pronounced suppression of NK activity in the spleen was found.

Herberman *et al.* (16) have determined the effect of host challenge on NK activity and found that in nude, as well as in normal mice of various ages, reactivity could be augmented by inoculation of tumor cells. This augmented cytotoxicity reached a peak three days after inoculation, and was only seen with tumor cells that bore target structures recognized by NK cells. The authors concluded that, in most respects, this augmentation of NK reactivity was consistent with the stimulation of specific memory cells by *in vivo* reexposure to antigen.

It has also been established by the same authors as well as by other groups that NK activity in mice could be augmented by other agents, notably allogenic cells, bacteria, and viruses. Wolfe and collaborators (17) demonstrated that viable BCG organisms given i.p. induced in the peritoneal cavity of normal mice a population of cytotoxic cells. These cells appeared within a few days after BCG administration, and had many features in common with NK cells. Also, i.p. administration by heat-killed *Corynebacterium parvum* elicited a similar phenomenon (Ojo *et al.*, personal communication). Thus, augmented NK activity after administration of bacterial adjuvants or viruses show that NK cells can be boosted nonspecifically, but the mechanism responsible for this nonspecific boosting is not known.

GENETIC REGULATION OF NK ACTIVITY

Early in the course of the study of NK cells in mice, it was found in different laboratories that various mouse strains differed consistently in NK reactivity (1, 2). This pronounced strain difference is indicative of a genetic influence and was investigated by further genetic analyses. High reactivity appears to be dominant; when the low reactive A-strain was crossed with various other strains, and cells from these F_1 hybrids tested for NK activity against semisyngeneic YAC lymphoma of A origin, reactivity resembled that of the high-reactive parent (18). This high NK activity in F_1 mice has suggested a possible relationship between NK cells and the so-called "hybrid effect." Since the time when Snell (19) first described this phenomenon, it has been shown in several tumor host systems that F_1 hybrid animals are generally more resistant to tumor growth of transplanted parental tumor cells than the strain of tumor origin. The precise mechanism operative in F_1 hybrid resistance remains obscure. Could NK cells represent this protective mechanism? Support for this notion derives from the fact that both phenomena are influenced by factors in the H-2 complex. The role of H-2 linked factors in the NK system was first shown in the YAC tumor system (20). For this purpose a linkage study in a backcross between the low-reactive A strain and a high-reactive F_1 hybrid was used. Highly significant linkage was established only with regard to the H-2 marker but no strong linkage was found with nine other markers segregating in this cross. Recent results confirmed the influence of H-2 linked factors on NK activity (8). Heterozygosity within or near the H-2D region was sufficient to ensure *in vivo* resistance as well as high NK reactivity *in vitro* against the EL-4 lymphoma.

References

1. Kiessling, R., Klein, E., and Wigzell, H. *Eur. J. Immunol.* 5, 112, 1975.
2. Herberman, R. B., Nunn, M. E., and Lavrin, D. H. *Int. J. Cancer* 16, 216, 1975.
3. Zarling, J. M., Nowinski, R. C., and Bach, F. H. *Proc. Nat. Acad. Sci. USA* 72, 2780, 1975.
4. Sendo, F., Aoki, T., Boyse, E. A., and Buofo, C. K. *J. Natl. Cancer Inst.* 55, 603, 1975.
5. Haller, O., Kiessling, R., Orn, A., Karre, K., Nilsson, K., and Wigzell, H. *Int. J. Cancer* 20, 93, 1977.
6. Becker, S., and Klein, E. *Eur. J. Immunol.* 6, 892, 1976.
7. Kiessling, R., Hochman, P., Haller, O., Shearer, G., Wigzell, H., and Cudkowicz, G. *Eur. J. Immunol.* 7, 655, 1977.
8. Harmon, R. C., Clark, E., O'Toole, C., and Wicker, L. *Immunogenetics* 4, 601, 1977.
9. Kiessling, R., Klein, E., Pross, H., and Wigzell, H. *Eur. J. Immunol.* 5, 117, 1975.
10. Herberman, R. B., Nunn, M. E., Holden, H. T., and Lavrin, K. H. *Int. J. Cancer* 16, 230, 1975.

11. Herberman, R. B., Nunn, M. E., Holden, H. T., Staal, S., and Djeu, J. Y. *Int. J. Cancer* **19**, 555, 1977.
12. Nunn, M. E., Djeu, J. Y., Glaser, J., Lavrin, D. H., and Herberman, R. B. *J. Natl. Cancer Inst.* **56**, 393, 1976.
13. Shellam, G. R. *Int. J. Cancer* **19**, 225, 1977.
14. Kiessling, R., Petranyi, G., Klein, G., and Wigzell, H. *Int. J. Cancer* **17**, 275, 1976.
15. Haller, O., Hansson, M., Kiessling, R., and Wigzell, H. *Nature* **270**, 609, 1977.
16. Herberman, R. B., Bartram, S., Haskill, S., Nunn, M., Holden, H., and West, W. *J. Immunol.* **119**, 322, 1977.
17. Wolfe, S. A., Tracey, D. E., and Henney, C. S. *Nature* **262**, 584, 1976.
18. Petranyi, G., Kiessling, R., and Klein, G. *Immunogenetics* **2**, 53, 1975.
19. Snell, G. D. *J. Natl. Cancer Inst.* **21**, 843, 1958.
20. Petranyi, G., Kiessling, R., Povey, S., Klein, G., Herzenberg, L., and Wigzell, H. *Immunogenetics* **3**, 15, 1976.

GENERATION *IN VIVO* OF MOUSE NATURAL CYTOTOXIC CELLS

Otto Haller

Natural cell-mediated cytotoxicity has recently been recognized as a potentially important effector mechanism common to several species including mice and man. It is now firmly established that lymphoid cells from nonimmune donors can specifically kill various kinds of tumor-derived or normal target cells *in vitro* as measured in short-term ^{51}Cr-release assays (1-3). At this stage of the investigations there has not yet emerged a comprehensive view of the *in vivo* relevance of natural cytotoxicity. As the mouse is a convenient species for *in vivo* manipulation, the generation and maturation of mouse natural killer cells (NK) was explored using various experimental approaches. The knowledge gained from such studies has led to the conclusion that mouse NK cells do participate in providing resistance to tumor cell growth *in vivo*.

GENERATION OF MOUSE NK CELLS *IN VIVO*

The notable features of the mouse NK system reported by most investigators are as follows: NK activity (a) shows distinct organ distribution, (b) is largely independent of a functional thymus, and (c) changes with the age of the individual mouse in a manner quite different from B- and T-lymphocyte function. Furthermore, NK cell activity is under strict genetic control with classification of inbred mouse strains as NK "high" or "low" reactive (3, 4, see also review by Kiessling in this volume). Accordingly, we have further investigated the generation, maturation, and maintenance of NK activity under controlled conditions *in vivo*.

7

A. Generation of NK Cells in Splenectomized Hosts

NK cells occur predominantly in spleen and blood of young adult mice. To define more precisely the role of the spleen in NK cell physiology, the effect of splenectomy on NK activity in high-reactive CBA/J mice was determined. Peripheral blood lymphocytes obtained four weeks after splenectomy or sham operation were assayed for *in vitro* NK activity on YAC-1 target cells, an *in vitro* line derived from a Moloney virus-induced lymphoma in A/Sn mice now widely used as a suitable target cell line for NK cell-mediated lysis. NK activity was retained in mice splenectomized as young adults. Bone marrow grafts restored NK activity in the spleen of irradiated histocompatible recipients (5). To ascertain whether an intact splenic microenvironment is required for this restoration, young adult CBA/J mice were splenectomized four weeks prior to lethal irradiation and syngeneic bone marrow reconstitution. As seen in Fig. 1, splenectomized animals were reconstituted to the same degree as control mice. Accordingly, the spleen is not mandatory for NK generation. These findings are further

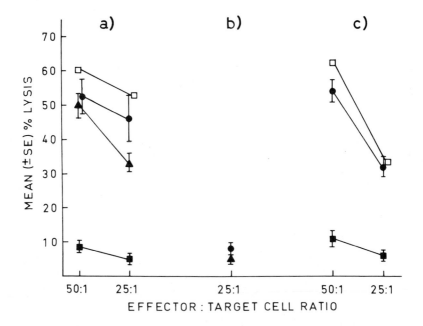

FIG. 1. Generation of NK cells in splenectomized hosts. Four weeks after splenectomy (▲——▲) or sham operation (●——●) CBA/J mice were irradiated with 800 R and reconstituted with syngeneic bone marrow cells. Eight weeks thereafter peripheral (a) blood lymphocytes, (b) lymph node cells, and (c) control spleen cells were tested individually for lytic activity against YAC-1. Each group consisted of a total of 10 mice. Additional controls comprised five untreated A/SW mice being low in NK activity (20) (■——■) and pooled effector cells from three normal CBA/J animals (□——□). Mean percent lysis ± SE is shown. For experimental details see ref. (6).

supported by the observation that lymph node cells from a mutant mouse lacking both spleen and thymus express relatively high NK activity (2).

B. Generation of NK Cells in Bone Marrow Chimeras

The generation of NK cells in the mouse has previously been shown to be under genetic control (4). The mechanisms through which NK activity is governed are largely unknown. Neither spleen nor thymus seem to be involved in the differentiation/maturation process leading to NK cells. Still, expression of NK activity might depend primarily on host environmental factors varying in different mouse strains. Thus, NK cells might gain specific cytotoxicity as a result of exposure and sensitization to ubiquitous antigens present in the host environment. Alternatively, the degree of NK cell activity might be genetically preprogrammed at the NK precursor cell level. To elucidate which factors were decisive for the generation of low or high NK activity *in vivo*, bone marrow chimeras between histocompatible, genetically high- or low-reactive mouse strains were produced. As summarized in Table 1, an exchange of bone marrow cells between mice sharing the same H-2 locus but differing in their degree of NK expression is followed by a restoration of spleen NK activity in irradiated recipients according to NK cell levels found in the spleens of the hemopoietic cell donors. Similar results were obtained with fetal liver cells transplanted instead of marrow cells. It would thus seem that the *in vivo* generation of NK cells is an inborn and autonomous function controlled by stem cells present in marrow or fetal liver. NK cell generation would seem not to depend on the genotype or other influences of the host environment.

C. Impairment of NK Cell Generation *in Vivo*

NK cell function is obviously preprogrammed at the level of NK precursor cells in the bone marrow. Any means of selectively disturbing the bone marrow compartment of the body might therefore be expected to interfere with NK cell activity as measured in peripheral organs. Indeed, ^{89}Sr treatment of young adult mice of high-NK genotype resulted in a significant reduction of spleen NK activity without similarly affecting other killer cell functions (7). Furthermore, *in vivo* administration of heteroantiserum against mouse bone marrow cells raised in rabbits significantly suppressed NK activity (8). However, NK activity levels were also reduced after *in vivo* treatment of mice with the antimacrophage agent silica known not to affect NK function when used *in vitro* (8). Thus, silica sensitive cells seem to be involved in the generation or maturation of NK cells *in vivo*. The differentiation pathway leading from precursor cells to mature NK cells could well be complex and awaits further clarification.

TABLE 1. NK Reactivity in Spleens from Bone Marrow Chimeras[a]

H-2 genotype	Bone marrow donors		Recipient animals[b]		NK activity of reconstituted chimeras[c]
	Strain	NK type[d]	Strain	NK type[d]	
b	C57L	high	C57L	high	high
	A.BY	low	C57L	high	low
	A.BY	low	A.BY	low	low
	C57L	high	A.BY	low	high
a	B10.A	high	A/Sn	low	high
	A/Sn	low	B10.4	high	low
a/b	(A/Sn x 129/J)F$_1$	low	(A/Sn x C57BL/6) F$_1$	high	low
	(A/Sn x C57BL/6)F$_1$	high	(A/Sn x 129/J)F$_1$	low	high

[a] Schematic summary, for detailed data see refs. (5) and (21).

[b] Eight to ten week old animals were lethally irradiated (750-850 R) and reconstituted intravenously with 3×10^7 viable bone marrow cells from sex-matched donors.

[c] Six to eight weeks after reconstitution spleen cells were assayed *in vitro* against ^{51}Cr-YAC-1 target cells.

[d] As assessed *in vitro* using spleen effector cells (22).

D. Induction of NK Cells *in Vivo*

The conditions leading to increase or decrease of NK activity in mice are complex. Consistent augmentation has been reported upon inoculation of mice with allogeneic cells either malignant or normal, with bacteria and with viruses (9). Intraperitoneal administration of viable BCG induces in the peritoneal cavity of normal mice a population of cytolytic effector cells displaying features common to NK cells (10, 11). We have observed a similar phenomenon following intraperitoneal inoculation of mice with heat-killed *Corynebacterium parvum* (*C. parvum*) (12). The peritoneal cavity of the mouse does not usually contain measurable numbers of NK cells. Figure 2 shows that cytolytic peritoneal excudate cells (PEC) appear early after inoculation of *C. parvum*. These cells are induced over a wide dose range and clearly differ from macrophages; they are indistinguishable from the NK cells found in normal adult spleen. Table 2 shows that the cytolytic specificity of *C. parvum*-induced PEC and the specificity pattern of splenic NK cells parallel one another when both kinds of effector cells are assayed against a panel of tumor target cell lines. Further investigations are required to clarify how these findings relate to the *in vivo* tumor growth inhibition observed upon treatment with bacterial immunostimulants and the possible contribution of NK cells to *in vivo* tumor resistance. It is possible that local tumor resistance

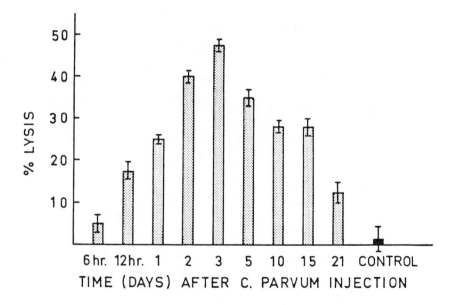

FIG. 2.   Time course of the appearance of cytolytic PEC following intraperitoneal inoculation of C parvum. PEC were harvested from groups of C3H/He mice that had been treated with 100 μg *C. parvum* i.p. for varying periods of time. These *C. parvum*-induced PEC were assayed together with normal control PEC from untreated animals against ^{51}Cr labeled YAC-1 target cells for 4 hr at an effector:target cell ratio of 50:1.

TABLE 2.   Specificity of Killer Cell Activity: Comparison Between C. Parvum Induced
PEC and NK Cells in Normal Spleen[a]

| Target cells[b] | % Specific target cell lysis[c] | | | |
| | Spleen NK cells[d] | | C. parvum-induced PEC[e] | |
	25:1	12.5:1	25:1	12.5:1
YAC-1	50.0	38.0	51.3	43.0
MOLT-4	23.5	10.8	31.1	19.0
P815	6.6	4.7	10.0	4.5
AKR-529	4.1	2.5	6.4	4.8

[a] For further details see ref (24).
[b] YAC-1 is a Moloney lymphoma of A/Sn origin (22), MOLT-4 is a human T-cell leukemia
susceptible to lysis by mouse NK cells (23), P815 of DBA/2 origin exhibits low susceptibility to NK cell mediated lysis (22), AKR-529 is a spontaneous AKR thymoma (12).
[c] Percent specific lysis was assessed *in vitro* for 4 hr at two different effector: target cell
ratios as indicated.
[d] Normal spleen cells from 6 to 8 weeks old CBA/H mice.
[e] Peritoneal exudate cells were obtained from groups of 6 to 8 week old CBA/H mice 3
days after intraperitoneal administration of 100 μg C. parvum.

induced by various agents could be mediated at least in part by NK cells, representing a hitherto neglected component of the host response induced by conventional adjuvants.

Role of NK Cells in Resistance Against Tumor Cell Growth *in Vivo*

In vivo resistance against tumor cell growth has previously been attributed
mainly to the action of thymus-dependent lymphocytes. T cells have generally
been considered to provide a major contribution to measurable resistance against
transplanted tumor cells (13). Yet, T-cell-deficient mice have frequently been
found to resist transplantation of various syngeneic, allogeneic, or xenogeneic
tumor grafts with unexpected vigor (14-19). Moreover, T-cell-deprived mice are

TABLE 3.   In Vivo Resistance Toward Semisyngeneic Tumor Cells in NK Chimeras

| Recipient animals[a] | Bone marrow donor[b] | NK type[e] | Frequency of tumor takes[e] | | |
			Day 7	Day 18	Day 29
(A/Sn x C57BL/6)F$_1$	(A/Sn x C57Bl/6)F$_1$	high	0/10	6/10	6/10
(A/Sn x C57BL/6)F$_1$	(A/Sn x A.BY)F$_1$	low	0/10	8/10	10/10
A/Sn Controls		low	0/5	5/5	5/5

[a] Irradiated with 750 R and reconstituted with anti-Thy and C-treated bone marrow cells.
[b] C57BL/6 and A.BY have the same H-2 = H-2B.
[c] NK activity of spleen cells from reconstituted mice as assessed in *in vitro*.
[d] Tumor challenge consisted of 10^4 cells of the YAC *in vivo* line (= H-2A) transplanted
subcutaneously. Number of animals with tumor growth over total number grafted. All
tumor-bearing mice died with tumor.

known to express high levels of NK activity (2, 3). Then, too, NK cells, several other T-cell-independent effector mechanisms comprising macrophages, or K cells (mediating antibody-dependent cellular cytotoxicity) could also be active in such animals. The relative contributions of these various effector cells for *in vivo* tumor protection prove difficult to assess. Preliminary evidence for a tumor-protective role of NK cells stems from experiments where a positive correlation has been established between the degree of NK activity (measured *in vitro*) and

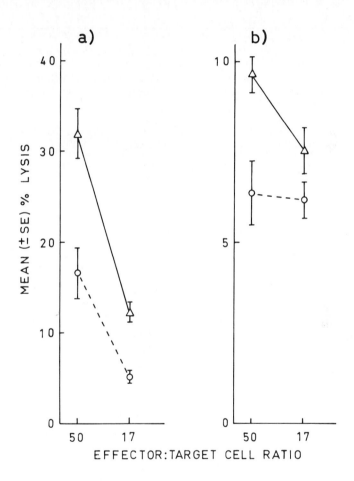

FIG. 3. Killer activity of spleen cells from F_1 hybrid mice against the in vitro line YAC-1 or the in vivo line of the YAC tumor. In parallel experiments individual spleens were tested at two different effector to target cell ratios against the *in vitro* line YAC-1 (a), and the *in vivo* YAC tumor line (b). Mean percent lysis ± SE is presented. △——△, (A/Sn x C57BL/6)F_1 effector cells. ○——○, (A/Sn x 129/J)F_1 effector cells. Each group consisted of four mice. At the 50:1 ratio, the differences in target cell killing observed between (A/Sn x C57BL/6)F_1 spleen cells and (A/Sn x 129/J)F_1 cells are significant at $p < 0.02$ for both target cell lines tested.

the expression of resistance in the same animal towards NK-sensitive tumor cells (18-20). In order to make this association more conclusive, we have looked for means to minimize the complexity of possible effector mechanisms usually involved in *in vivo* phenomena. Adult A/Sn F_1 hybrid mice were irradiated and reconstituted with marrow cells from histocompatible high or low NK donors (see Table 3). The NK activity of the spleen cells from such chimeras is predetermined at the level of stem cells transplanted from the bone marrow donor. Thus, for the first time, hosts were available for *in vivo* tumor challenge that expressed

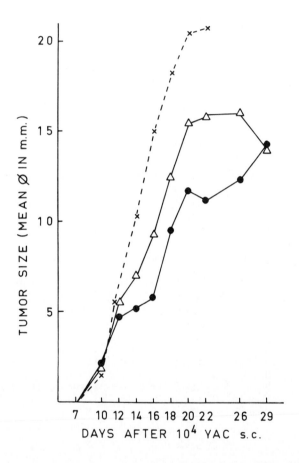

FIG. 4. NK chimeras: tumor size increae in tumor-bearing animals. NK chimeras were the same as listed in Table 3. Subcutaneous tumor challenge dose was 10^4 cells of the YAC *in vivo* line. The mean tumor diameter of all animals developing subcutaneous tumors in each group was determined. x——x, A/Sn control mice (five out of five animals with tumors). ●, (A/Sn x C57BL/6)F_1 mice irradiated with 750 R and reconstituted with (A/Sn x C57BL/6)F_1 anti-Thy serum + C-treated bone marrow cells (six out of ten animals developed tumors). △(A/Sn x C57BL/6)F_1 mice irradiated with 750 R and reconstituted with anti-Thy serum + C-treated bone marrow cells (ten out of ten animals developed tumors). For detailed conditions of transfer see ref. (21).

either high or low NK activity (as measured *in vitro*) and yet were genetically near identical. Groups of chimeric and control mice were challenged subcutaneously with small tumor inocula using an *in vivo* line of the YAC tumor. As is shown in Fig. 3, the *in vivo* line is quite sensitive to NK activity although at a considerably lower level than the *in vitro* line YAC-1. Incidence of tumor takes and mean tumor diameters were recorded. Table 3 shows a good correlation between *in vivo* tumor resistance and expected *in vitro* NK activity, in that the high NK F_1 hybrids displayed better resistance than the NK low hybrids when transplanted with 10^4 YAC cells. Additional experiments with animals thymectomized before irradiation and marrow protecting yielded essentially the same results (21). Obviously in the present system T lymphocytes are not essential for tumor resistance. It should be noted, however, that the tumor resistance observed was not absolute. When the tumor load was increased tenfold, all mice died from tumor outgrowth . Furthermore, all animals that developed tumors when transplanted with 10^4 YAC cells showed the same rate of increase in tumor size and, when they finally succumbed, there were only minor differences in survival time (Fig. 4).

NK activity changes with the age of the individual mouse, appearing three weeks after birth and declining at around three months of age. In our NK system, young animals at peak NK activity were found to be more resistant to subcu-

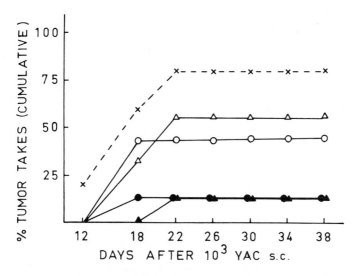

FIG. 5. In vivo resistance toward semisyngeneic tumor cells in animals of different ages. (A/Sn x CBA)F_1 or (A/Sn x C57BL/6)F_1 mice of different ages were inoculated subcutaneously with 10^3 YAC cells (*in vivo* line). Percentage tumor takes was recorded. Both of these F_1 combinations are high in NK activity (20). NK low reactive (A/Sn mice, 4 months of age) were included as controls. Each group consisted of seven to nine mice. All animals carrying tumors died with tumors. A/Sn controls: x —— x; (A/Sn x CBA)F_1 mice: ▲—— ▲, 3½ weeks old; △—— △, 8 months old; (A/Sn x C57BL/6)F_1 mice: ●—— ●, 3½ weeks old; ○ —— ○, 6 months old.

taneous growth of small numbers of YAC lymphoma cells than were old mice (Fig. 5). *In vivo* tumor resistance would thus seem to parallel the age-related changes of *in vitro* NK activity. Our findings are consistent with previous results on age-related changes of *in vivo* resistance toward a radiation-induced lymphoma (18).

These data thus support the view that naturally occurring killer cells play a decisive part in resistance against certain syngeneic tumor cells *in vivo*. This protective role has so far been shown for only a limited number of tumors. Further work is required to ascertain the general validity of these findings. However, recent evidence is sufficient to suggest the existence of an alternative, nonthymus-dependent surveillance mechanism to the conventional immune system.

Acknowledgments

This work has been supported by the Swiss Academy for Medical Research, the U. S. National Cancer Institute (contract N01-CB-64033), and the Swiss National Science Foundation (grant 3.139-0.77).

References

1. Haller, O., Kiessling, R., Gidlund, M., and Wigzell, H. "Proceedings of the International Symposium on Tumor-Associated Antigens and Their Specific Immune Response," F. Spreatico and R. Arnon, eds. Academic Press, New York, 1979.
2. Herberman, B., and Holden, T. *Adv. Cancer Res.* 27, 305-377, 1978.
3. Kiessling, R., and Haller, O. *Contemp. Top. Immunobiol.* 8, 171-201, 1978.
4. Petranyi, G. G., Kiessling, R., and Klein, G. *Immunogenetics* 2, 53-61, 1975.
5. Haller, O., Kiessling, R., Oern, A., and Wigzell, H. *J. Exp. Med.* 145, 1411-1416, 1977.
6. Haller, O., Gidlund, M., Kurnick, J. T., and Wigzell, H. *Scand. J. Immunol.* 8, 207-213, 1978.
7. Haller, O., and Wigzell, H. *J. Immunol.* 118, 1503-1506, 1977.
8. Kiessling, R., Hochman, P. S., Haller, O., Shearer, G. M., Wigzell, H., and Cudkowicz, G. *Eur. J. Immunol.* 7, 655-663, 1977.
9. Herberman, R. B., Nunn, M. E., Holden, H. T., Staal, S., and Djeu, J. *Int. J. Cancer* 19, 555-564, 1977.
10. Tracey, D. E., Wolfe, S. A., Durdik, J. M., and Henney, C. S. *J. Immunol.* 119, 1145-1151, 1977.
11. Wolfe, S. A., Tracey, D. S., and Henney, C. S. *J. Immunol.* 119, 1152-1158, 1977.
12. Ojo, E., Haller, O., and Wigzell, H. *Scand. J. Immunol.* 8, 215-222, 1978.
13. Cerottini, J. C., and Brunner, K. T. *Adv. Immunol.* 18, 67-132, 1974.
14. Epstein, A. L., Herman, M. M., Kim, H., Dorfman, R. F., and Kaplan, H. S. *Cancer* 37, 2158-2176, 1976.
15. Gillette, R. W., and Fox, A. *Cell Immunol.* 19, 328-335, 1975.
16. Giovanella, B. C., Stehlin, J. S., and Williams, L. J., Jr. *J. Natl. Cancer Inst.* 52, 921-930, 1974.
17. Maguire, H., Jr., Outzen, H. C., Custer, R. P., and Prehn, R. T. *J. Natl. Cancer Inst.* 57, 439-442, 1976.

18. Sendo, F., Aoki, T., Boyse, E. A., and Buafo, C. K. *J. Natl. Cancer Inst.* 55, 603-609, 1975.
19. Warner, N. L., Woodruff, M. F. A., and Burton, R. C. *Int. J. Cancer* 20, 146-155, 1977.
20. Kiessling, R., Petranyi, G., Klein, G., and Wigzell, H. *Int. J. Cancer* 17, 1-7, 1976.
21. Haller, O., Hansson, M., Kiessling, R., and Wigzell, H. *Nature* 270, 609-611, 1977.
22. Kiessling, R., Klein, E., and Wigzell, H. *Eur. J. Immunol.* 5, 112-117, 1975.
23. Haller, O., Kiessling, R., Oern, A., Karre, K., Nilsson, K., and Wigzell, H. *Int. J. Cancer* 20, 93-105, 1977.
24. Ojo, E., Haller, O., Kimura, A., and Wigzell, H. *Int. J. Cancer* 21, 444-452, 1978.

ARE CYTOTOXIC T CELLS RELEVANT IN THE LOCAL DEFENSE AGAINST SOLID TUMORS?

Stephen Haskill and Susanne Becker

Spontaneous regression of human tumors is a most unusual occurrence. Although spontaneous regression of subclinical tumors is believed to occur, such a phenomenon is in fact difficult, if not impossible to prove. There are various animal models, which provide evidence for the immunological control of spontaneous and drug-induced regression. The responsible mechanisms appear to depend upon thymus-derived cells but the actual role in the process attributed to these cells is still not identified. In particular, our knowledge of the effector mechanisms present and active in local defense against solid tumors is as yet sketchy. Based on both morphology and function, various host-infiltrating-cell types have been identified in a number of solid tumors [see ref. (1) for review]. Thymus-derived lymphocytes, monocytes, and macrophages as well as natural killer cells, have all been detected in both human and animal tumors. However, the experimental data are insufficient to draw a conclusion as to the relevant roles of the various kinds of effector cells in host defense against the growing solid tumor. It is the purpose of this communication to outline the results obtained in two distinct murine spontaneous regression models in which appreciable knowledge of the *in situ* effector mechanisms is available. The presence of thymus-derived lymphocytes in these solid tumors is readily demonstrated; however, in neither system has it been possible to detect cytotoxic T cells *active* against the appropriate *in situ* target cells.

19

T1699 TUMOR MODEL

The T1699 mammary adenocarcinoma is a readily transplantable murine adenocarcinoma originally derived from a spontaneous mammary tumor that developed
in a DBA/2 mouse. The site of subcutaneous inoculation regulates the ultimate
fate of the transplanted cells. Spontaneous regression occurs if 5×10^4 to 10^5
cells are inoculated subcutaneously in the abdominal area, whereas progression
always results if the same number of cells are inoculated subcutaneously in the
shoulder fatpad area (2). Chemotherapy of progressing tumors results in a regression dependent on the immune status of the host (3). Spontaneous regression is also under immune control as sublethally irradiated or ALG-treated hosts
support progressive growth of the tumor, even with drug therapy (3, 4). Both
regressor and progressor tumors are infiltrated by thymus-derived lymphocytes,
eosinophils, FcR-bearing monocytes, and large highly phagocytic macrophages
(Fig. 1). All attempts to demonstrate cytotoxic T-cell activity with either peripheral lymphoid cells or the infiltrating cells have been negative. The test assays
have been chromium and proline release, colony inhibition, microcytotoxicity
employing periods of incubation up to three days and using effector cells from

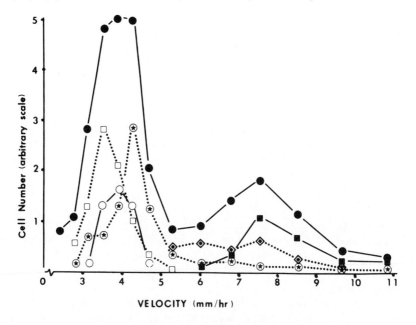

FIG. 1.   Sedimentation Velocity Profile of 14-day T1699 Regressor Tumor. 14 day regressor tumors were dispersed in 0.14% collagenase-DNase solution and separated at unit gravity
by sedimentation velocity [see ref. (2)]. The host fraction of infiltrating cells (Hfn) includes
all cells sedimenting between 3 and 5.2 mm/hr. Large phagocytic macrophages were obtained
by adherence of cells sedimenting between 7 and 10 mm/hr. ● ●, Total cells; □ □, Theta-
positive cells; ○ ○, eosinophils; ⊛ ⊛, EA rosettes; ◆ ◆, phagocytes; ■ ■, tumor cells.

the lymph nodes, Peyer's patches, spleen, and blood (2). In contrast to the cytotoxicity data, however, when the adoptive transfer assay for delayed hypersensitivity is used, thymus-derived effector cells can be detected in both the spleen
and the tumor (4). As cytotoxic T-cell responses could not be detected in this
model and yet immune-dependent spontaneous and drug-induced regression
occurred, a search was made for an alternate mechanism of tumor inhibition.
Unlike MSV-induced tumors (5, 6) and rat fibrosarcomas (7, 8), the proportion
of infiltrating cells with the morphology of activated macrophages was extremely
low. Therefore nonspecific growth inhibition by activated macrophages was un-

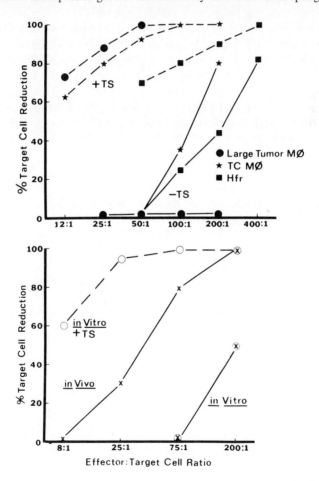

FIG. 2. Microcytotoxicity test for antibody-dependent cellular cytotoxicity (ADCC) using
tissue culture macrophages (Mφ) grown from bone marrow (9), the Hfn of infiltrating cells
(see Fig. 1) and large phagocytic macrophages from the tumor (Fig. 1). (a) ADCC activity
against tissue culture T1699 target cell. (b) Tumor-derived large macrophage activity against
T1699 cells ± tumor serum or against *in vivo* T1699 cells isolated from the same tumors as
the macrophages—no antibody added [see also ref. (9)].

likely to be responsible for the regression (2).

Both progressor and regressor tumor bearers do, however, produce large amounts of antibody directed against T1699 membrane antigens. This immune response appears to be thymus dependent (9) and the antibody is active when tested in an ADCC assay with macrophages and monocytelike cells isolated from bone marrow, blood, and also from the tumor (Fig. 2a) [see also ref. (10)]. The active antibody belongs to the IgG_{2a} subclass (9) and appears earlier in the serum of regressors than progressors (4). None of the other classes and subclasses are functionally active against these target cells either when macrophages or lymphoid cells are used as effector cells. Not only are ADCC effector cells present in the tumor, but also the tumor cells are coated with antibody *in situ* (9). This has been shown by indirect fluorescence on *in situ* derived tumor cells, by elution of antibody from purified tumor cells at low pH, and by demonstrating that *in situ* tumor cells are more sensitive to macrophages *in vitro* than are tumor cells isolated from immunologically incompetent animals not producing antibodies. Because both effector and target cells can be isolated from the same solid tumors (Fig. 2b) the evidence suggests that an important mechanism of defense against the T1699 tumor is likely to be ADCC.

Further evidence comes from the more clinically relevant *in vivo* experiments involving drug-induced regression in immunocompetent mice. Radov *et al.* (3) have reported that the chemotherapeutic effectiveness in animals carrying progressor tumors is dependent on the immune status of the host. That is, immunosuppressed tumor-bearing animals fail to show drug-induced regression. If Melphalan is given prior to full establishment of the antibody response, all tumorbearing animals fail to respond to chemotherapy (11). However, if these early drug-treated and therefore immunosuppressed tumor-bearing animals are given tumor-specific antisera, the normal percentage of animals responding to the drug is restored (Table 1). This would seem to indicate that chemotherapy-induced regression may well require an intact T-cell dependent, macrophage-mediated ADCC.

MOLONEY SARCOMA TUMORS–A/Sn MICE

The Moloney sarcoma virus-induced tumor system is another interesting model showing spontaneous regression, provided immunocompetent mice are used for tumor induction. Alterations of thymic function leads to tumor progression (12-15). Transfer of immune lymphoid cells to T-cell deficient animals results in regressing tumors, while pretreatment of the immune cells with antitheta serum and complement abolishes the protective effect (16), thus clearly demonstrating the important role of T cells in the host response to this tumor.

In a number of laboratories the nature of effector cells present in primary MSV-induced tumors has been investigated. Cytotoxic T cells (5, 17), growth inhibitory macrophages (5), and NK (18-19) cells have been isolated from these

TABLE 1. Synergism between Early Chemotherapy and
Passive Transfer of Immune Serum

Treatment	Regression index[a]
None	0/20
100 μg Melphalan D 6	0/20
0.5 ml anti-T1699 Serum days 7-12	0/20
0.5 ml anti-CaD$_2$ [b] Serum days 7-12	0/20
0.5 ml anti-T1699 Serum days 7-12 100 μg Melphalan D 6	14/20
0.5 ml anti-CaD$_2$ Serum days 7-12 100 μg Melphalan D 6	0/20

[a] All animals injected with 10^5 T1699 cells s.c. in the shoulder (Progressors).
[b] CaD$_2$ is noncross-reacting DBA-2 mammary adenocarcinoma.

tumors. The importance of cytotoxic T cells in the elimination of the sarcoma cells has been assumed. However, the cytotoxic activity of the infiltrating T cells has only been demonstrated using MLV-induced lymphomas as targets, since they are believed to express the relevant target antigens (20). In A/Sn mice, however, Becker and Klein (18) were unable to detect cytotoxic T cells when syngeneic lymphomas were used as targets. Neither spleen nor infiltrating cells were active in cytotoxic T-cell tests. However, low levels of NK activity were found in the infiltrating population, suggesting a possible role of these cells in the control of tumor growth.

The indication that cytotoxic T cells might not be necessary for MSV tumor regression led us to reexamine the question of effector cells infiltrating the A/Sn MSV tumor. Autologous "tumor" cells could be separated from infiltrating cells in sufficient purity by sedimentation at unit gravity to provide suitable target cells for *in vitro* assessment of *in situ* immunity (Fig. 3). These target cells were demonstrated to possess abnormal chromosome numbers, and had morphologic features associated with tumor cells (Becker and Haskill, unpublished data). It has subsequently been demonstrated, however, that these cells fail to passage *in vitro* for extended periods of time and eventually result in cultures containing what appear to be normal fibroblasts. The target cells thus seem to be infected rather than transformed by the virus, a result entirely in keeping with the histologic studies in this tumor system (21, 22).

Using the effector cells isolated from the MSV tumors following fractionation by sedimentation at unit gravity, we were able to demonstrate that NK activity could be separated from the activity against the autologous target cells. NK acti-

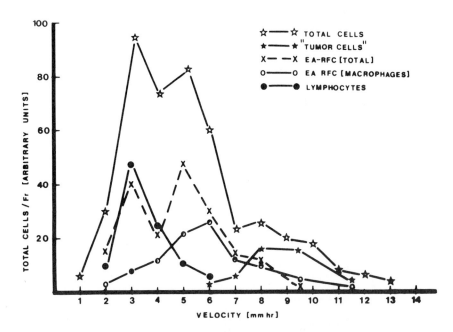

FIG. 3. Sedimentation velocity profile of collagenase-DNase dispersed MSV primary tumors (day 11). "Tumor" cells and lymphocytes were quantitated by morphology on Giemsa-stained cytocentrifuge preparations. SRBC coated with high dilutions of antibody were used to detect phagocytic macrophages (EARFC macrophage) and lower dilutions of antibody were used to coat SRBC for total FcR bearing cells (4). Infiltrating fraction 3-5 mm/hr contained 45% Θ positive cells.

vity was restricted to higher velocities, while cells active against the autologous targets were found in the lowest velocity fractions (Fig. 4). Thus, NK cells apparently had little influence on the survival of autologous target cells. The cell fractions active against the autologous "tumor" cells were treated with antitheta serum and complement. However, the cytotoxic potential in these fractions was not reduced by this treatment (Table 2). Thus, despite the fact that significant

TABLE 2. Effect of R Anti-BΘ + C' Treatment on Autologous Target Cell Reduction

	I	II	III
Normal Spleen	86.0 ± 9.0	N D	44.2 ± 1.8
Tumor spleen (C')	71.1* ± 4.8	29.2 ± 2.9	N D
Tumor spleen (BΘ + C')	N D	23.8 ± 2.2	N D
Infiltrating Fn (C')	52.8 ± 5.3	16.4 ± 0.9**	26.3 ± 2.5
Infiltrating Fn (BΘ + C')	49.0 ± 3.4	10.8 ± 2.6***	N D
Media	82.6 ± 3.3	39.9 ± 3.8	44.8 ± 3.2

[a] No complement treatment.

[b] 45% Θ positive cells.

[c] 4% Θ positive cells.

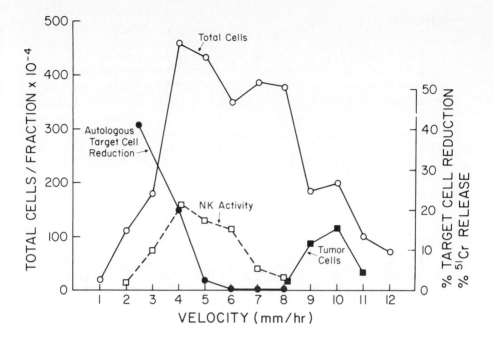

FIG. 4. Sedimentation velocity profile of both natural killer (NK) and autologous activity induced with various fractions of infiltrating cells present in regressor MSV tumors (see Fig. 3 also). NK assay was carried out with ^{51}Cr labeled YAC target cells as in ref. (18). Autologous activity was determined in the microcytotoxicity test against "tumor" cells obtained from the high-velocity fractions 10-13 mm/hr. NK assay carried out at 50:1 and autologous assay at 500:1 ratio of effector to target cells.

numbers of T cells were found in the infiltrating population, they apparently were inactive in the autologous target cell reduction assay. The results provided strong evidence that neither cytotoxic T cells *nor* NK cells were likely *in situ* effector cells against the MSV-infected cells found in these tumors.

The data for a variety of T-cell tests carried out using tumor infiltrating cells have been given elsewhere (1). These include cytotoxicity, delayed hypersensitivity mitogen response, and blast cell responses to autologous target cells. Of particular note are the early studies of Nairn *et al.* (23) and Nind *et al.* (24), who reported that in two human tumor systems studied peripheral blood cytotoxic lymphocytes could be detected in over 50% of the patients, but lymphocytes isolated from the same tumor as the target cells nonetheless failed to show activity against these same cells. There are several possible explanations for these findings. It could be argued that cytotoxic T cells were not detected because their receptors were blocked by the high levels of antigens presumably present inside the tumors. The fact that a variety of mitogen responses have been detected with the infiltrating T cells (1) and that in some animal systems delayed hypersensitivity T cells have been detected *in situ* argues against such an explanation.

CONCLUDING STATEMENT

A variety of defense mechanisms are demonstrable inside regressing and progressing solid tumors. The emphasis placed on the role of T cells in both spontaneous and drug-induced regression of tumors may not be entirely warranted. Without denying the potential role of cytotoxic lymphocytes *in situ*, it is affirmed that a variety of other mechanisms are involved in host defense.

Acknowledgments

This work was supported in part by the National Cancer Institute Contract No. NO1-CB-64023, grants from the Swedish Cancer Society; USPHS Grant CA-17694, and ACS Grant IM-84 (to S. H.). Part of the investigations were done during the tenure of S. H. at the Transplantation Laboratory, (Dr. P. Hayry) University of Helsinki, funded by a UICC, ICRTT Award.

References

1. Haskill, J. S., Hayry, P., and Radov, L. *Contemp. Top. Immunobiol.* 1977.
2. Haskill, J. S., Yamamura, Y., and Radov, L., *Int. J. Cancer* **16**, 798-809, 1975.
3. Radov, L., Haskill, J. S., and Korn, J. H. *Int. J. Cancer* **17**, 773-779, 1976.
4. Haskill, J. S., Radov, L., Yamamura, Y., Parthenais, E., Korn, J. H., and Ritter, F. L. *J. Reticuloendothelial Soc.* **20**, 233-241, 1976.
5. Holden, H. T., Haskill, J. S., Kirchner, H., and Herberman, R. B. *J. Immunol.* **117**, 440-446, 1976.
6. Russell, S. W., Gillespie, G. Y., and McIntosh, A. T. *J. Immunol.* **118**, 1574-1579, 1977.
7. Haskill, J. S., Proctor, J. W., and Yamamura, Y. *J. Natl. Cancer Inst.* **54**, 387-393, 1975.
8. Evans, R. Academic Press, New York, 1976.
9. Haskill, J. S., Radov, L. A., Fett, J. W., and Parthanais, E. *J. Immunol.* **119**, 1000-1006, 1977.
10. Haskill, J. S. *Int. J. Cancer* 1979.
11. Radov, L. A., Korn, J., and Haskill, J. S. *J. Immunol.* **18**, 630-638, 1976.
12. Ting, R. C. *Proc. Soc. Exp. Biol. Med.* **126**, 778-781, 1976.
13. East, J., and Harvey, J. J. *Int. J. Cancer* **3**, 614-627, 197 .
14. Law, L. W., Ting, R. C., and Allison, A. C. *Nature* **220**, 611-612, 197 .
15. Allison, A. C., Marga, J. N., and Hammond, V. *Nature* **252**, 746-747, 1974.
16. Gorczynski, R. M. *J. Immunology* **112**, 553-539, 1974.
17. Plata, F., MacDonald, H. R., and Sordat, B. *Bibl. Haemat.* **43**, 274-277, 1976.
18. Becker, S., and Klein, E. *Eur. J. Immunol.* **6**, 892-989, 1976.
19. Plata, F., and Sordat, B. *Int. J. Cancer* **19**, 205-211, 1977.
20. Levy, J. D., and Leclerc, J. C. *Adv. Cancer Res.* **24**, 1-66, 1977.

21. Stanton, M. F., Law, L. W., and Ting, R. C. *J. Natl. Cancer Inst.* **40**, 1113-1129, 1968.
22. Siegler, M. D. *J. Natl. Cancer Inst.* **45**, 135-147, 1970.
23. Nairn, R. C., Nind, A. P. P., Guli, E. P. G., Davies, D. J., Holland, J. M., McGiven, A. R., and Hughes, E. S. R. *Brit. Med. J.* **4**, 706-709, 1971.
24. Nind, A. P. P., Nairn, R. C., Rolland, J. W., Guli, E. P. G., and Hughes, E. S. R. *Brit. J. Cancer* **28**, 108-117, 1973.

NATURAL CYTOTOXICITY OF HUMAN LYMPHOCYTES. IMMUNOGLOBULIN DEPENDENT AND INDEPENDENT SYSTEMS

P. Perlmann, M. Troye, G. R. Pape, B. Harfast, and T. Andersson

Lymphocytes from immune donors are cytotoxic to target cells carrying the appropriate surface antigens. The effector lymphocytes operative in this cytotoxicity *in vitro* may be of different kinds. Thus, both *in vivo* and *in vitro* immunization to alloantigens of the major histocompatibility locus (MHC) gives rise to the generation of cytotoxic T lymphocytes (CTL) that do not require humoral antibodies for recognition of the target cells (1). CTL of this type are also generated by immunization with syngeneic cells, modified by virus or by haptenization (2, 3), with syngeneic tumor cells (4, 5) or with cells differing from the responder cells in minor histocompatibility antigens (6). The CTL-mediated cytotoxicity in these "altered self" systems is syngeneically restricted, i.e., the target cells and the effector cells must be identical in regard to some of their major histocompatibility antigens (7).

Since CTL-mediated cytotoxicity is independent of conventional antibody, it is not inhibited by addition of antiimmunoglobulin reagents. This latter inhibition is easily achieved when cytotoxicity is mediated by antibody-dependent effector cells (K cells), i.e., cells that require conventional humoral antibody as recognition factors for target cell antigens (1, 8). K-cell-mediated reactions are usually studied by adding antitarget cell antibodies to mixtures of target cells and lymphocytes from normal donors. However, antibody dependent K-cell-mediated reactions also take place when lymphocytes from immune donors are mixed with target cells without addition of external antibodies. In these instances, cytotoxicity has been shown to constitute a cell cooperation phenomenon in which the antibody-producing cells and the effector cells are distinct (9).

More recently, it has been noted that lymphocytes from apparently normal donors are often cytotoxic *in vitro* to a variety of target cells. This phenomenon is called "natural" or "spontaneous" cytotoxicity. Although natural cytotoxicity may reflect an undiscovered sensitization of the lymphocyte donor and hence be due to one of the mechanisms discussed above, there is also evidence for a natural cytotoxicity that does not fit into any of these schemes. This latter type of cytotoxicity has been most extensively studied in mice and the corresponding effector cells have been designated NK cells (natural killer cells) (10, 11). These cells appear to be lymphocytic cells that lack known T- or B-cell markers as well as monocyte/macrophage characteristics. NK activity in different mouse strains is genetically determined and in some as yet undefined manner linked to the MHC system. Although NK reactivity appears to be selective, i.e., it does not affect target cells indiscriminately, the structural basis of its specificity is unknown and its possible immunological nature has not been established. However, NK cells are believed to be responsible for an innate surveillance system against certain tumors.

A natural cytotoxicity of an apparently similar kind has also been found in man. However, in the systems studied thus far the human effector cells appear to have both Fc-receptors for IgG and certain surface markers typical for T cells, e.g., low-avidity receptors for sheep erythrocytes (11). This implies that the effector cells mediating this natural cytotoxicity are very similar to human K cells (12). Therefore, it is possible that natural cytotoxicity in these instances reflects an antibody-dependent reaction, perhaps mediated by natural antibodies produced by the lymphocyte donor. In the following, we will describe experiments that were designed to further elucidate the possible immunoglobulin dependency of natural cytotoxicity in different human systems.

TABLE 1. Cytotoxicity of Lymphocytes from Healthy Donors

	% ^{51}Cr-Release Above Or Below Background[a]							
	T24[b]		HCV29[b]		HT29[b]		MEL1[b]	
Donor	60:1	30:1	60:1	30:1	60:1	30:1	60:1	30:1
1	18.8	12.8	32.2	18.1	–	–	–	–
2	6.2	−0.4	2.2	−1.9	–	–	–	–
3	4.8	13.8	15.7	10.3	–	–	–	–
4	3.6	3.3	1.7	−0.9	−1.6	−0.7	11.1	6.9
5	1.1	2.0	6.9	8.7	−6.9	−8.7	–	–
6	−0.6	−2.5	–	–	–	–	20.6	9.2
7	−2.7	−7.8	3.0	–	–	–	1.0	–
8	−3.1	−2.9	1.4	0.7	7.6	5.4	–	–
9	−3.4	−13.9	6.7	0.3	18.4	–	–	–
10	−6.4	−5.3	30.2	22.6	–	–	–	–

[a] Corrected by subtracting % release from labeled target cells in lymphocyte-free medium controls after 18 hr of incubation, lymphocyte: target cell ratios 60:1 and 30:1, respectively.

[b] All target cells were from established cell lines. T24, transitional cell carcinoma (TCC) of urinary bladder; HCV29, bladder epithelium; HT29, colon carcinoma; MEL1, malignant melanoma. For details see (13).

NATURAL CYTOTOXICITY DEPENDENT ON THE PRESENCE
OF IMMUNOGLOBULIN

Table 1 shows the typical reaction of lymphocytes from 10 healthy donors, assayed for cytotoxicity against a small panel of allogeneic target cells of different histogenetic origin. All four target cell types were from established cell lines and were of about equal susceptibility to lymphocyte-mediated cytotoxicity when tested in the ^{51}Cr release assay (13). In most instances, the lymphocytes were not or only weakly cytotoxic. However, occasionally, cytotoxicity was strong and significant. Moroeever, when occurring, it was usually selective (14), i.e., lymphocytes from individual donors were cytotoxic against one or sometimes two of the target cells on the panel but not against all of them. The selectivity of this natural cytotoxicity suggested that it had a true immunological basis, involving a sensitization of the lymphocyte donors to one or several antigens on the target cells that were lysed (13).

For further studies we selected lymphocytes from cancer patients and healthy donors, displaying a significant cytotoxicity against some of the target cells shown in Table 1. Fractionation of these purified blood lymphocytes, surface marker analysis, and cytotoxicity experiments showed that the effector cells responsible for natural cytotoxicity could not be separated from K cells (15, 16). They had Fc receptors for IgG but were clearly distinct from B cells as characterized by surface bound IgG (SIg). A significant fraction (but not all) of the effector cells in both natural cytotoxicity and a K-cell model also had sheep erythrocyte receptors and may thus be T cells. In order to establish the possible immunoglobulin dependency of the reactions, we prepared Fab fragments of rabbit antibodies to human IgG (Fab-aHuIg). Fab fragments were chosen when it was found that variable results were obtained with bivalent $F(ab')_2$ fragments (17). Because of the occasional occurrence in rabbit serum of natural antibodies to human cells, all antibodies were purified on unsolubilized human IgG. For control, we used Fab fragments of immunoadsorbent purified rabbit antibodies to ovalbumin (Fab-aOA).

The Fab-aHuIg reagents completely inhibited the antibody-dependent K-cell-mediated reactions against tumor cells induced by the addition of alloantibodies. A typical example is shown in Fig. 1. In contrast, no inhibition was obtained when the reagent was added to a cytotoxic system in which the effector cells were antibody-independent CTL, i.e., to a cell-mediated lympholysis system (CML) induced by MLC activation in vitro (17).

Subsequently, the natural cytotoxicity of both selected normal donors and cancer patients to the tumor cells shown in Table 1 was investigated in the same manner. When the Fab-aHuIg reagent was added to the incubation mixtures, a dose-dependent inhibition of cytotoxicity was seen in all instances. However, the intensity of this inhibition varied for individual donors. Only rarely, 100% inhibition was achieved. In most instances, inhibition leveled off at about 50-60% and could not be increased by further increasing the dose of inhibitor. Addition of

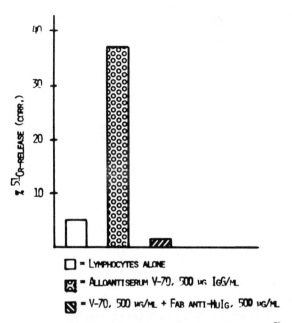

FIG. 1. Inhibition of antibody-dependent cytotoxicity by Fab-aHuIg. ^{51}Cr-labeled target cells (HT29, colon carcinoma cell line) were incubated as indicated with lymphocytes alone, with lymphocytes in the presence of 500 μg IgG/ml of an alloantiserum (V-70) or in the presence of both V-70 and Fab-aHuIg (500 μg/ml): Time of incubation 18 hr, lymphocyte: target cell ratio 30:1. The ordinate shows % ^{51}Cr-release corrected by subtracting % release in lymphocyte-free medium control.

Fab-aOA gave no inhibition in any case. Experiments with immunoadsorbent purified reagents directed either against the F(ab')$_2$ fragments or against the Fc fragments of human IgG were both inhibitory. Reagents directed to Fc structures of the human IgG subclasses were also inhibitory. However, while antibodies to Fc from IgG1 were inhibitory in almost all instances, antibodies to Fc from IgG2, IgG3, and IgG4 inhibited the cytotoxicity of normal donors' lymphocytes less frequently. Preliminary experiments with Fab-anti-IgM have thus far given negative results (17).

Although SIg$^+$ B cells are not required for the natural cytotoxicity discussed above (16), the possibility remained that the inhibitory effects of Fab-aHuIg on cytotoxicity were due to the interaction of the reagent with B cells (or rather, with cells carrying adsorbed IgG). In order to investigate this we fractionated the lymphocytes on antiimmunoglobulin columns that depleted them of SIg$^+$ cells while ∿75% of the lymphocytes with Fc receptors for IgG (FcR) were left behind in the passed fraction (15). Table 2 shows the typical results with the cells of a healthy donor whose lymphocytes were cytotoxic to three different target cell types. It is seen that removal of SIg$^+$ cells only led to a weak reduction of natural cytotoxicity. However, this was strongly inhibited by Fab-aHuIg, regardless of whether SIg$^+$ cells were present or not.

TABLE 2. Effect of Fab-Antihuman IgG on Natural Cytotoxicity of Human Lymphocytes

Target Cells[a]	Lymphocytes[b]	% ^{51}Cr-Release[c]	% Inhibition[d]
T24	U	47.5	78
	F	39.2	75
MANO	U	62.7	63
	F	52.5	65
MEL-1	U	38.0	61
	F	32.6	66

[a] T24, MANO, TCC of urinary bladder; MEL-1, malignant melanoma.
[b] From normal donor MK; U, unfractionated; F, B-cell depleted by passage through $F(ab')_2$-anti HuIg column; see (15).
[c] Corrected by subtracting release from labeled target cells in lymphocyte-free medium controls after 18 hr of incubation; lymphocyte: target cell ratios 60:1.
[d] % Inhibition of ^{51}Cr-release in presence of Fab-aHuIg, 50 μg/ml incubation mixture.

Our experiments suggest that a considerable fraction of the natural cytotoxicity of lymphocytes from almost all normal donors against the allogeneic target cells used in these studies was not only dependent on FcR^+ effector cells but also on the presence of IgG. Similar results were obtained with the disease-related cytotoxicity of cancer patients. The findings lend support to the notion that natural cytotoxicity in these instances reflects, at least in part, an antibody-dependent K-cell reaction, in line with the findings reported by others using completely different methods (18). If so, it also appears that the immunoglobulin involved does not originate from the SIg^+ cells present in the blood. However, the blood lymphocyte preparations always contain some cells that are SIg^- but which contain cytoplasmatic immunoglobulin and are not removed on the antiimmunoglobulin columns (G. Pape, unpublished). Therefore, it is possible that cytotoxicity is induced by immunoglobulin released from these cells during the 18 hr of incubation with the target cells. Alternatively, cytotoxicity could be due to cytophilic immunoglobulin carried over into the incubation mixtures with the lymphocytes. However, if so, this cytophilic immunoglobulin would have to be present in a form that makes it unavailable for binding to the antiimmunoglobulin columns. Moreover, no Fab-aHuIg inhibition of cytotoxicity has been seen in preliminary experiments in which the lymphocytes were pretreated with the reagents before they were added to the target cells. Further experiments are needed to establish the origin of the inducing immunoglobulin.

NATURAL CYTOTOXICITY NOT DEPENDENT ON IMMUNOGLOBULIN

Although these results suggest that the immunoglobulin dependency of natural lymphocyte-mediated cytotoxicity is a very frequent phenomenon, we do not imply that it is an exclusive one. As pointed out above (17), addition of Fab-aHuIg, even at high concentrations did not usually lead to complete inhibition of natural cytotoxicity. It remains to be established whether the incomplete inhibition seen in many experiments had some technical reasons or whether it reflected a heterogeneity of the underlying effector mechanisms. Evidence for an immunoglobulin- and FcR^+-effector-cell-independent cytotoxicity against the tumor cells under consideration has been discussed elsewhere (16; and Troye *et al.*, this volume).

An additional example of natural lymphocyte cytotoxicity independent of IgG was recently found in experiments with virus-infected target cells. It has previously been reported that lymphocytes from most normal donors are cytotoxic to target cells acutely infected with Parotis virus (19). This cytotoxicity has a considerable degree of cross reactivity and is not syngeneically restricted, implying that the effector cells are not of the CTL type. Since they were also FcR^+ and were distinct from the majority of the lymphocytes with receptors for sheep erythrocytes (20), this natural cytotoxicity would have been expected to reflect an antibody-dependent K-cell-mediated reaction as has been seen in other human virus systems where the lymphocytes were obtained from vaccinated donors (21). However, when tested in the presence of Fab-aHuIg, the natural cytotoxicity of normal donors' lymphocytes to mumps virus-infected target cells was not inhibited (22). A typical example is shown in Table 3. No inhibition was seen in other experiments in which Fab-aHuIg was added at even higher concentrations (up to 170 μg/ml) (22). It should be mentioned that the same reagent inhibited the natural cytotoxicity of some of the donors to the same target cells before virus infection. Moreover, it also completely inhibited cytotoxicity against mumps virus-infected cells by normal lymphocytes when it was induced by addition of anti-mumps IgG from patients with Parotis infection (22). The mechanism of this natural cytotoxicity to virus-infected target cells is presently unknown. However,

TABLE 3.   Effect of Fab-Antihuman IgG or Control Anti-Body on Natural Cytotoxicity of Human Lymphocytes to Mumps Virus-Infected Target Cells

Dose[a] (μg/ml)	^{51}Cr-Release[b]	
	Fab-aHuIg	Fab-aOA
0	29	29
2	28	32
8	31	30
32	28	24

[a] Fab-aHuIg or Fab-aOA present in incubation mixture at concentration indicated.

[b] From mumps virus-infected Vero (green monkey kidney) cells labeled with ^{51}Cr, corrected by subtracting % release in medium controls. Lymphocytes from one healthy donor, lymphocyte: target cell ratio 60:1, time of incubation 18 hr. For details see (22).

it resembles the enhanced natural cytotoxicity displayed by some FcR$^+$ human lymphocytes when confronted with interferon-inducing agents such as viruses or certain "inducer" cell lines, as recently described by Trinchieri *et al.* (23, 24).

CONCLUDING REMARKS

The findings discussed in this paper illustrate the fact that natural cytotoxicity of human lymphocytes from normal donors or certain patients is a heterogenous phenomenon. We have presented evidence that it frequently depends on FcR$^+$ lymphocytes that display cytotoxicity in conjunction with immunoglobulin (IgG). We assume that the inducing immunoglobulin is natural antibody to target cell antigens, i.e., that cytotoxicity in these instances represents a K-cell-mediated reaction. However, this assumption remains to be formally proven. In addition, we have shown that natural cytotoxicity in some cases may be independent of immunoglobulin, despite the fact that the effector cells involved are very similar to the antibody-dependent K cells. This may mean that the same FcR$^+$ effector cells can be induced to become cytotoxic by different triggering mechanisms. However, it is also possible that cells belonging to different lymphocyte subsets are involved in the different systems. In any event, our results indicate that the net cytotoxicity of an individual donor's lymphocytes may be the result of several distinct reactions occurring simultaneously. It is obvious that lymphocyte fractionation and surface marker analysis alone are not always sufficient to clarify the mechanisms of a cytotoxic reaction.

Immunoadsorbent-purified antibodies to human immunoglobulin, applied as Fab fragments, appear to be powerful tools for establishing the involvement of immunoglobulin in some of these reactions.

Acknowledgments

This work was supported by grant no. B78-16X-00148 from the Swedish Medical Research Council and by a NCI contract CB-74129. G. R. Pape is holder of a fellowship from Deutsche Forschungsgemeinschaft.

References

1. Cerottini, J. C., and Brunner, T. K. *Adv. Immunol.* 18, 67, 1974.
2. Doherty, P. C., Blanden, R. V., and Zinkernagel, R. M. *Transplant. Rev.* 29, 89, 1976.
3. Shearer, G. M., Rehn, T. C., and Schmitt-Verhulst, A. M. *Transplant. Rev.* 29, 222, 1976.
4. Plata, F. Jongeneel, V., Cerottini, J. C., and Brunner, K. T. *Eur. J. Immunol.* 6, 823, 1976.

5. Trinchieri, G., Aden, D. P., and Knowles, B. B. *Nature* **261**, 312, 1976.
6. Gordon, R. D., Simpson, E., and Samuelson, L. E. *J. Exp. Med.* **142**, 1108, 1975.
7. Zinkernagel, R. M., and Doherty, P. C. *Contemp. Top. Immunobiol.* **7**, 779, 1977
8. Perlmann, P. *In* "Clinical Immunobiology" (F. Bach and R. Goods, eds.)., Vol. 3, p. 107. Academic Press, New York, 1976.
9. Schirrmacher, V., Rubin, B., and Pross, H. *J. Immunol.* **112**, 2219, 1974.
10. Kiessling, R., and Haller, O. *Contemp. Top. Immunobiol.* **8**, 171, 1978.
11. Herberman, R. B., and Holden, H. T. *Adv. Cancer Res.* **27**, 305, 1978.
12. Perlmann, P., Perlmann, H., Wåhlin, B., and Hammarström, S. Quantitation, fractionation, and surface marker analysis of IgG- and IgM-dependent cytolytic lymphocytes (K-cells) in human blood. *In* "Immunopathology 7th International Symposium 1976" (P. A. Miescher, ed.), p. 321. Schwabes, Basel, 1977.
13. Troye, M., Perlmann, P., Larsson, Å., Blomgren, H., and Johansson, B. *J. Cancer* **20**, 188, 1977.
14. Bean, M., Bloom, B., Herberman, R., Old, L., Oettgen, H., Klein, G., and Terry, W. *Cancer Res.* **35**, 2902, 1975.
15. Pape, G. R., Troye, M., and Perlmann, P. *J. Immunol.* **118**, 1919, 1977.
16. Pape, G. R., Troye, M., and Perlmann, P. *J. Immunol.* **118**, 1925, 1977.
17. Troye, M., Perlmann, P., Pape, G. R., Spiegelberg, H. L., Näslund, I., and Gidlof, A. *J. Immunol.* **119**, 1061, 1977.
18. Akira, D., and Takasugi, M. *Int. J. Cancer* **19**, 74, 1977.
19. Andersson, T., Stejskal, V., and Härfast, B. *J. Immunol.* **114**, 237, 1975.
20. Härfast, B., Andersson, T., and Perlmann, P. *J. Immunol.* **114**, 1820, 1975.
21. Perrin, L. M., Zinkernagel, R. H., and Oldstone, M. B. A. *J. Exp. Med.* **146**, 949, 1977.
22. Härfast, B., Andersson, B., and Perlmann, P. **127**, 755, 1978.
23. Trinchieri, G., Santoli, D., and Koprowski, H. *J. Immunol.* **120**, 1869, 1978.
24. Trinchieri, G., and Santoli, D. *J. Exp. Med.* **147**, 1314, 1978.

CHARACTERIZATION OF EFFECTOR CELLS IN HUMAN NATURAL CYTOTOXICITY

Gerd R. Pape, Marita Troye, and Peter Perlmann

Natural cytotoxicity of human blood lymphocytes against target cells from the myeloid cell line K 562 was analyzed in a ^{51}Cr release assay. The effector cells involved were characterized and some data favoring the role of immunoglobulin in the induction of this cytotoxicity are presented.

METHODS

Lymphocytic effector cells isolated from heparinized peripheral blood by gelatin sedimentation, carbonyl iron treatment, and Ficoll-Isopaque centrifugation were characterized by an analysis of the following surface markers: Surface-bound immunoglobulin (SIg$^+$) assayed by direct immunofluorescense; receptors for sheep erythrocytes (E$^+$) and for the Fc part of IgG (EAG$^+$), assayed by rosette formation as described elsewhere (1); and the receptor for the carbohydrate-binding A hemagglutinin from the snail *Helix pomatia* (HP) by direct immunofluorescence with FITC-conjugated HP (2). After treatment with neuraminidase, 60-80% of the lymphocytes in human peripheral blood display HP receptors. HP receptors are present on the majority of the SIg$^-$E$^+$ lymphocytes and the HP receptor is considered to be a T-cell marker for adult human blood lymphocytes. The effector cells were fractioned on columns as follows:

(1) B cells (SIg$^+$ cells) but not K cells were removed on glass bead columns

with antiimmunoglobulin activity but lacking IgG-Fc structures [F(ab')$_2$/anti-F(ab')$_2$ columns].

(2) Cells with Fc receptors for IgG were removed on glass bead columns charged with immune complexes (ovalbumin/antiovalbumin columns) (3).

(3) HP fractionation was performed by passing neuraminidase-treated lymphocytes through a column charged with Sepharose to which HP was bound covalently (2).

Lymphocyates without HP receptors pass directly through the column (fraction I). Retained HP$^+$ cells are then eluted stepwise with medium containing the competitive sugar N-acetyl-D-galactosamine at a concentration of 0.1 mg/ml (fraction II) and 1.0 mg/ml (fraction III), respectively. Fraction I contains the majority of the B cells (SIg$^+$ cells), a large part of the cells bearing Fc receptors for IgG but only a minor part of the HP receptor bearing cells. Fraction II is of intermediate composition in regard to surface marker distribution. The majority of the T cells with HP receptors is eluted in fraction III, which usually contains 90-95% HP$^+$ cells (2). The natural cytotoxicity of the lymphocyte subpopulations was analyzed in a short term ^{51}Cr-release assay (4 hr of incubation) with target cells from the myeloid cell line K 562 (4). 5 x 10^3 labeled target cells were incubated in each tube at different lymphocyte:target cell ratios. The IgG-dependent K-cell potential of the lymphocytes was determined in an ADCC model system, either with sensitized ^{51}Cr labeled bovine erythrocytes as target cells (1) or with sensitized ^{51}Cr-labeled tumor cells, preferentially T24, a cell line derived from a transitional cell carcinoma of the human urinary bladder (5). The sensitizing antibody was from rabbits immunized with T24 (M. Schneider, unpublished). Fab fragments of rabbit IgG antibodies to human immunoglobulin (IgG) (or to ovalbumin as control) were prepared and purified on immunoadsorbents as described elsewhere. The fragments were added to the ^{51}Cr-release assay at a concentration of 100 μg/ml, previously shown to give optimal inhibition under the present experimental conditions (6).

RESULTS AND DISCUSSION

The natural cytoxicity of lymphocytes from two normal donors after fractionation through F(ab')$_2$/anti-F(ab')$_2$ columns or through ovalbumin/antiovalbumin (OA/anti-OA) columns is presented in Table 1. Removal of B (SIg$^+$) cells on F(ab')$_2$/anti-F(ab')$_2$ columns led to no significant reduction of NK activity. On the other hand, removal of Fc receptor bearing cells by OA/anti-OA fractionation resulted in conspicuous loss of natural cytotoxicity. The K cell activity against T24 of the fractionated subpopulations was likewise abolished (not shown in Table 1). Table 2 provides a typical example of NK activity of lymphocytic subpopulations fractionated on HP columns. Neuramini-

TABLE 1. Cytotoxic Effects to K 562 Target Cells of Column Fractionated Lymphocytes

	% ^{51}Cr-release from K 562 target cells[a]					
	Unfractionated Lymphocytes		F(ab')$_2$/anti-F(ab')$_2$[b] fractionated lymph.		OA/anti-OA[c] fractionated lymph.	
Donor	60:1[d]	30:1	60:1	30:1	60:1	30:1
M.S.	51.7	46.7	45.5	31.4	4.3	1.5
S.H.	46.5	33.0	46.7	33.5	8.4	3.0

[a] [51]Cr release corrected by subtracting background release in medium controls (5%). Incubation time: 4 hr.

[b] Columns charged with F(ab')$_2$ fragments of human IgG and F(ab')$_2$ fragments of rabbit antibodies to F(ab')$_2$ of human IgG.

[c] Columns charged with ovalbumin and rabbit antibodies to ovalbumin.

[d] Lymphocyte:target cell ratios 60:1 or 30:1.

dase (NANAse) treatment did not significantly change the NK potential. In the experiment of Table 2 the natural cytotoxicity of the treated lymphocytes was slightly enhanced. The NK activity in fraction I was also slightly enhanced as compared with that of unfractionated cells. It was clearly enhanced in fraction II but always reduced in fraction III. However, residual NK activity in fraction III was still significant. It was abolished by passing these cells through an OA/ anti-OA column (data not shown), a procedure known to remove the HP[+]-EAG[+] cells present in this fraction (2). In general, the distribution of NK activity among the three fractions paralleled that of ADCC seen in model systems and was also well correlated with the distribution of Fc receptor bearing lymphocytes (2; and unpublished data).

Overall, the experiments demonstrate that effector cells responsible for NK activity against the myeloid cell line K 562 belong to the same subpopulation of peripheral blood cells that display K cell activity. They are nonadherent, nonphagocytic, SIg[-] cells bearing Fc receptors for IgG. A significant fraction of NK cells belongs to the T-cell lineage as characterized by E and HP receptors. This characterization of NK cells is further supported by fractionation of the

TABLE 2. Cytotoxic Effects to K 562 Target Cells of HP-Column Fractionated Lymphocytes

	% ^{51}Cr-Release[a]									
					Lymphocytes					
	Untreated		NANAse treated[b]		HP I[c]		HP II		HP III	
Donor	60:1[d]	30:1	60:1	30:1	60:1	30:1	60:1	30:1	60:1	30:1
M.T.	34.9	19.7	39.7	24.6	42.6	29.2	61.6	44.5	16.3	8.2

[a] [51]Cr release from K 562 target cells (see legend to Table I).

[b] Lymphocytes were incubated with neuraminidase (10 μg/ml) Cl. perfringens, type VI, 1-3 U/mg, NAN-Lactose substrate (Sigma Chem. Col., St. Louis, Missouri).

[c] Lymphocyte fractions obtained from the HP column (see text).

[d] Lymphocyte:target cell ratio.

TABLE 3. Inhibition of Natural Cytotoxicity Against K 562 by Fab-Fragments
of Antihuman Immunoglobulin

	Lymphocyte : Target cell ratio			
	60:1		30:1	
Donor	% ^{51}Cr Release[a]	% Inhibition[b]	% ^{51}Cr Release	% Inhibition
M.S. ♂	51.7	13.7	46.7	36.8
M.T. ♀	34.9	22.3	19.7	n.d.
S.P. ♂	36.1	62.0	28.3	64.6
L.M. ♂	50.3	n.d.	45.1	95.1
L.B. ♀	12.0	80.0	8.8	75.0
F.B. ♂	21.8	63.7	14.1	51.7
S.H. ♂	46.5	33.7	33.0	43.6
B.H. ♂	47.2	61.6	30.9	69.5
C.M. ♀	52.4	63.1	38.1	64.8

[a] ^{51}Cr release from K 562 target cells (see legend to Table 1).
[b] % inhibition of corrected ^{51}Cr release from K 562 target cells by addition of Fab fragments of immunoadsorbent purified rabbit antibodies to human IgG at a concentration of 100 μg/ml.

lymphocytes first into E$^+$ and E$^-$ cells and subsequently into EAG$^+$ and EAG$^-$ cells by rosetting with bovine erythrocytes coated with rabbit IgG (7; and manuscript in preparation). The presence of Fc receptors on natural killer cells has been described recently (8, 9). The present results are in agreement with our previous data demonstrating the involvement of Fc receptor bearing lymphocytes in the spontaneous cytotoxicity against various tumor lines in culture (10). The NK cells against K 562 are different from the NK cells of mice described by Kiessling et al. (11, 12). According to these authors, murine NK cells lack B- and T-cell markers as well as Fc receptors and monocyte/macrophage characteristics. In contrast, Herberman et al. recently gave evidence for the presence of Fc receptors for IgG on mouse NK cells (13).

NK cells could not be separated from K cells by our fractionation procedures. To investigate the role of immunoglobulin in the cytotoxic reaction against K 562, purified Fab fragments of antihuman Ig were added to the assay system. As is evident in Table 3, natural cytotoxicity was inhibited to a different extent for individual donors when Fab anti-Ig was added at a concentration previously shown to yield maximal inhibition (6). The degree of inhibition ranged from 14 to 95%. Fab fragments of anti-OA antibodies were used for control in all experiments. No inhibition was observed after addition of these fragments (data not shown). After removal of B (SIg$^+$) cells from the effector cell population, the inhibitory effect of Fab antiimmunoglobulin was unchanged (Table 4). The NK activity of the different fractions obtained by HP column fractionation was also inhibited by Fab anti-Ig (Table 5). Furthermore, pretreatment of lymphocytes with trypsin strongly reduced their NK activity. This was not restored when the lymphocytes were subsequently incubated with cycloheximide (20 μg/ml). These data (not shown) suggest a requirement of a trypsin-sensitive structure

TABLE 4. Effect of Fab/Anti-Ig on Cytotoxicity of Lymphocytes Against K 562 After Depletion of B (SIg⁺) Cells

Lymphocyte fraction[a]	Ratio[b]	Antibodies added (100 µg/ml)[c]	% Corrected ⁵¹Cr-Release[d]
Unfractionated	60:1	none	46.5
		Fab/anti-Ig	30.8
		Fab/anti-OA	46.9
	30:1	none	33.0
		Fab/anti-Ig	18.6
		Fab/anti-OA	35.4
B-Cell depleted	60:1	none	46.7
		Fab/anti-Ig	25.1
		Fab/anti-OA	48.0
	30:1	none	33.5
		Fab/anti-Ig	17.3
		Fab/anti-OA	36.7

[a] B-cell depleted fraction obtained by passage of the lymphocytes through $F(ab')_2$/anti-$F(ab')_2$ column.
[b] Lymphocyte:target cell ratio.
[c] Addition of Fab fragments of immunoadsorbent purified rabbit antibodies to human IgG or to ovalbumin, respectively.
[d] ⁵¹Cr release from K 562 target cells (see legend to Table 1).

TABLE 5. Effect of Fab-Fragments of Antihuman Immunoglobulin on the Cytotoxicity Against K 562 Target Cells of Lymphocyte Subpopulations After *Helix Pomatia* Column Fractionation

Lymphocyte fractions	Ratio[a]	% ⁵¹Cr Release[b]	% Inhibition by addition of Fab/a-Ig[c]
Unfractionated	60:1	20.8	49.5
	30:1	14.2	38.0
HPI	60:1	25.3	55.7
	30:1	20.4	46.5
HPII	60:1	29.6	53.7
	30:1	19.3	40.4
HPIII	60:1	12.5	33.6
	30:1	9.9	32.3

[a] Lymphocyte to target cell ratio.
[b] ⁵¹Cr release from K562 target cells (see legend to Table 1).
[c] % inhibition of corrected ⁵¹Cr release (see legend to Table 3).

for the cytotoxic reaction, which is dependent on protein synthesis. The trypsin-sensitive structure could well be cytophilic antibody on the lymphocyte surface but other interpretations are possible.

These results suggest that K cells are the effector cells and that a very significant part of the natural cytotoxicity against the myeloid cell line K 562 is mediated by Ig antibody. Whether cytophilic antibodies rather than those released from Ig-producing cells (during the cytotoxic assay) play a major role in the present system remains to be determined. It is emphasized that both the B-cell depleted fractions and all three HP-column fractions invariably contained cells with cytoplasmic Ig. Such cells could be responsible for the release of the inducing Ig. On the other hand our experiments also show that not all of this natural cytotoxicity was inhibitable by the anti-Ig reagent. Thus, in addition to a K-cell mechanism, other Ig-independent mechanisms may be operative.

Acknowledgments

This work was supported by grant no. B 78-16X-00148 from the Swedish Medical Research Council. G. R. Pape is holder of a fellowship from Deutsche Forschungsgemeinschaft.

References

1. Perlmann, H., Perlmann, P., Pape, G. R., and Halldén, G. *Scand. J. Immunol.* 5 (Suppl. 5), 57, 1976.
2. Hellström, U., Hammarstrom, S., Dillner, M.-L., Perlmann, H., and Perlmann, P. *Scand. J. Immunol.* 5 (Suppl. 5), 45, 1976.
3. Pape, G. R., Troye, M., and Perlmann, P. *J. Immunol.* 118, 1919, 1977.
4. Lozzio, C. B., and Lozzio, B. B. *J. Natl. Cancer Inst.* 50, 535, 1973.
5. Troye, M., Perlmann, P., Larsson, A., Blomgren, H., and Johansson, B. *Int. J. Cancer* 20, 188, 1977.
6. Troye, M., Perlmann, P., Pape, G. R., Spiegelberg, H. L., Naslund, I., and Gidlof, A. *J. Immunol.* 119, 1061, 1977.
7. Moretta, L., Ferrarini, M., Mingari, M. C., Moretta, A., and Webb, S. R. *J. Immunol.* 117, 2171, 1976.
8. Jondal, M., and Pross, H. *Int. J. Cancer* 15, 596, 1975.
9. West, W. H., Cannon, G. B., Kay, H. D., Bonnard, G. D., and Herberman, R. B. *J. Immunol.* 118, 355, 1977.
10. Pape, G. R., Troye, M., and Perlmann, P. *J. Immunol.* 118, 1925, 1977.
11. Kiessling, R., Klein, E., and Wigzell, H. *Eur. J. Immunol.* 5, 112, 1975.
12. Kiessling, R., Klein, E., Pross, H., and Wigzell, H. *Eur. J. Immunol.* 5, 117, 1975.
13. Herberman, R. B., Bartram, S., Haskill, J. S., Nunn, M., Holden, H. T., and West, W. H. *J. Immunol.* 119, 322, 1977.

EFFECTOR CELLS OF NATURAL CYTOTOXICITY
AGAINST HUMAN MELANOMA CELLS

E. P. Rieber, J. G. Saal, M. Hadam, G. Riethmüller,
H. Rodt, and M. P. Dierich

In a number of species including man, a lymphocyte population has been described that displays spontaneous cytotoxicity *in vitro* against syngeneic and allogenic tumor cells but without previous sensitization of the lymphocyte donor. This phenomenon, also called natural cytotoxicity, has been ascribed to effector cells characterized by the expression of Fc receptors but lacking the typical markers of either T or B lymphocytes (1-3). A similar "null" profile of surface characteristics was originally drawn for the K cells that are responsible for antibody-dependent cell-mediated cytotoxicity. Recent reports, however, implied that T marker bearing lymphocytes were also acting as effector cells in both natural cytotoxicity and antibody-dependent cellular cytotoxicity (1, 4-7). Since in these two mechanisms allogeneic restriction is not observed, as imposed on lysis by immune T cells, it was of interest to study further the nature of the effector cells involved in natural cytotoxicity. The expression of Fc receptors on the surface of these cells led to the question of whether by analogy with K cells humoral antibodies are utilized as specific recognition molecules (6, 8-10).

In the present study effector cells in spontaneous cytotoxicity were analyzed for their T-cell nature by use of a heterologous antithymus antiserum and E-rosette formation. Furthermore, purified Fab-antihuman Ig reagents were used to demonstrate the presence of Fc receptor-bound immunoglobulin on these cells and to evaluate the functional involvement in the cytolytic reaction of the receptor-immunoglobulin complex.

43

MATERIALS AND METHODS

Target Cells

Two cell lines derived from human malignant melanoma and established in our laboratory were used as target cells: Mel-Im-82 and Mel-Ei-78. The cells were cultured in RPMI 1640 (Gibco) supplemented exclusively with 10% human AB-serum, antibiotics, and L-glutamine.

[3]H-Proline Release Test

CMC was measured by the ^{3}H-proline release test as previously described in detail (11). 5×10^7 tumor cells were labeled for 24 hr with 50 μCi ^{3}H-proline (500-1000 mCi/mmol) and distributed into Falcon Microtest II Plates (1×10^4 per well). Effector lymphocytes at various concentrations were added in RPMI 1640 supplemented with 10% human AB-serum.

ADCC was assessed by adding a human antimelanoma antiserum to the CMC system at a final dilution of 1:6. ADCC was defined as antibody-induced cytolysis superimposed on spontaneous CMC.

After 12 hr (ADCC) and 40 Hr (CMC) of incubation at 37°C, the total contents of each well were collected on glass filters by use of an automatic harvesting device. The radioactivity retained on the filter was determined in a liquid scintillation counter. The cytolytic activity (CTL) of lymphocytes was calculated by

$$\% \text{ CTL} = 100 - \frac{\text{cpm test sample}}{\text{cpm medium control}} \times 100$$

Cytolytic activity of lymphocytes was finally expressed as lytic units, one lytic unit (LU) being defined as the number of lymphocytes required to achieve 25% CTL.

Lymphocytotoxic Test

The IgG fraction of a highly specific rabbit antihuman thymocyte serum (ATCG) was used to identify and quantitate human T-cell antigen-bearing lymphocytes (12). The cells to be tested were suspended in RPMI 1640 with 10% heat-inactivated human AB serum and finally adjusted to a concentration of 1.25×10^7 cells/ml. An equal volume of 1:4 diluted ATCG was added and the cells were incubated for 30 min at room temperature. Selected, normal undiluted rabbit serum (final concentration 71.4%) served as a source of complement in a second incubation step of 90 min at room temperature.

Lymphocyte Purification

Nonadherent lymphocytes (NAL) from normal male individuals were isolated from peripheral blood by Ficoll^R Urovison^R (density: 1080 gm/ml) density gradient centrifugation. Adherent cells were removed by incubation on plastic Petri dishes and by passage of the supernatant cells through a nylon wool column as previously described (11). Determination of surface markers on lymphocytes was performed as described (6, 11).

Separation of Lymphocyte Subpopulations

(a) Depletion of complement receptor lymphocytes (CRL). Lymphocytes bearing receptors for either C3b or C3d were identified by the method of Dierich *et al.* (13) using sheep-red blood cells (SRBC) for preparation of 19 S-EAC3b/3d. Equal volumes of lymphocytes at 5×10^6 /ml in RPMI 1640 and 19S-EAC3b (or 19S EAC3d) at 2.5×10^8 /ml in RPMI 1640 were incubated (30 min, 37°C), centrifuged (5 min, 37°C), incubated again (10 min, 37°C), resuspended, and layered on Ficoll^R Urovision^R (density: 1.140 gm/ml). CRL was depleted by centrifugation at 400 x g for 30 min at 37°C. The whole procedure was performed at 37°C in order to avoid rosette formation via the SRBC receptor of T-cells.

(b) Depletion of lymphocytes bearing Rc receptors. Lymphocytes with receptors for the Fc portion of IgG were removed by passage of NAL through IgG-anti-IgG bead columns as previously described (6).

(c) Separation of lymphocytes forming rosettes with SRBC. 1 ml of lymphocytes at 1×10^7 cells/ml in fetal calf serum (FCS) was mixed with 2 ml SRBC (2.5×10^8/ml) in FCS. This mixture was incubated at 37°C for 5 min, centrifuged at 58 x g for 5 min and incubated over night at 4°C. Following incubation, the pellet was resuspended, 3 ml of RPMI 1640 were added and the cells suspension was poured onto 20 ml Ficoll^R Urovison^R (density: 1.080 gm/ml). The gradient was centrifuged at 400 x g at room temperature for 40 min. For dissociation of SRBC from E-RFC, the cell pellet was resuspended in 20 ml prewarmed (37°C) RPMI 1640 supplemented with 50% FCS. The cell suspension was again layered onto prewarmed (37°C) Ficoll^R Urovison^R (density: 1.140 gm/ml). For separation of SRBC and lymphocytes, the cell suspension was centrifuged at 37°C and 400 x g for 40 min.

(d) Isolation of ATCG-resistant E-RFC. 2.5×10^7 E-RFC separated as described above in 500 $\mu\ell$ RPMI 1640 supplemented with 10% heat-inactivated human AB serum were mixed with 500 $\mu\ell$ of 1:4 diluted ATCG. The cells were then incubated for 30 min at room temperature and 2.5 ml undiluted selected rabbit serum as complement source were added. Thereafter, the cell suspension was further incubated at room temperature until the percentage of lysed cells remained constant (about 90 min). The removal of dead cells was performed by Ficoll^R Urovision^R centrifugation (density: 1.080 gm/ml) at room temperature.

Immunoradioautography

Lymphocytes were stained with either ^{125}I-Fab(ab)$_2$-antihuman γ or ^{125}I-F(ab)$'_2$-antihuman μ antibodies or ^{125}I-labeled aggregated IgG of defined size (6) for 30 min at 4°C, washed extensively, mixed with SRBC (100 SRBC per 1 lymphocyte), and incubated for an additional 2 hr at 4°C. Smears of the suspension were prepared on slides and fixed in methanol for 20 min. Further processing of the immunoautoradiographs was performed according to routine methods with the exception that development was done at 27°C and exposure time was at least 30 days. For evaluation of double marker cells, 400 cells were counted from each preparation.

RESULTS

Nonadherent lymphocytes (NAL) of healthy individuals were separated into E-rosette forming cells (E-RFC) and non-E-rosette forming cells (non-E-RFC) by density gradient centrifugation. When the recovered cell populations were tested

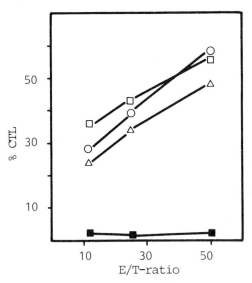

FIG. 1.   Spontaneous cytolytic activity of NAL (○), E-RFC (△), non-E-RFC (□), and IgG-anti-IgG column filtered lymphocytes (■). Cell fractions were prepared as described in "Materials and Methods" and tested simultaneously on Mel-Ei-78 target cells. Each point represents the mean value of six experiments.

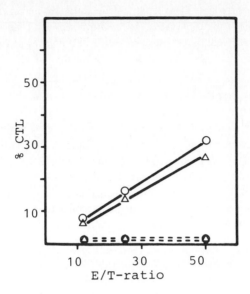

FIG. 2. Inhibition of spontaneous cytolytic activity of NAL (○) and E-RFC (△) by aggregated IgG. NAL an E-RFC (2 x 10⁶/ml) were preincubated with either RPMI 1640 medium ($\underset{\text{o}}{\text{o}} \!=\! \underset{\text{o}}{\text{o}}$) or aggregated human (gG (6mg/ml) ($\underset{\text{o}}{\text{o}} \!=\! =\! \underset{\text{o}}{\text{o}}$) for 30 min at 4°C prior to testing on Mel-Ei-78 target cells. Each point represents the mean value of three experiments.

for spontaneous cytolytic activity (CMC) against melanoma cells, E-RFC revealed a CMC usually lower than that of NAL. In contrast, non-E-RFC exhibited a slightly higher CMC than NAL per cell number (Fig. 1). When, however, CMC is calculated on the basis of recoveries of E-RFC [52.3% ± 1.9 (SD)] and non-E-RFC [9.7% ± 1.9 (SD)], about ¾ of the lytic units contained in the NAL fraction can be attributed to E-RFC.

As can be deduced from Fig. 1, spontaneous cellular cytotoxicity depends on the presence of Fc receptors (FcR) or s-Ig, since CMC is completely lost after passage of NAL through immune complex columns. The involvement of the Fc receptor in the cytolytic activity of isolated E-RFC could also be demonstrated using aggregated IgG as a probe for inhibition of FcR activity. As shown in Fig.2, addition of aggregated IgG resulted in a complete disappearance of cytolytic activity of both NAL and E-RFC.

To determine whether the cytolytic effector cells also bear receptors for C3b or C3d complement components, NAL were depleted of cells forming rosettes with either 19S-EAC3b or 19S-EAC3d. As shown in Table 1, NAL could be effectively depleted of cells carrying complement receptors. However, the remaining complement receptor-negative cells proved to be as cytotoxic as unfractionated cells at every effector cell-target cell ratio tested (Fig. 3). Identical results were obtained for C3b and C3d rosettes, but seemed to be crucially dependent on the use of purified IgM antibodies for the sensitization of SRBC. Minute amounts of contaminating IgG antibodies were found to lead to a decrease of

TABLE 1. Surface Marker Analysis

Lymphocyte preparations	S-Ig[b] bearing cells (%)	^{125}I-aggIgG binding cells (%)	E-RFC (%)	19S-EAC3b-RFC (%)	19S-EAC3d-RFC (%)
NAL unfractionated	7.1 ± 1.8	8.4 ± 0.9	85.6 ± 7.4	26.5 ± 0.7	—
NAL EAC3b-depleted	—	—	—	<0.5	—
NAL EAC3d-depleted	—	—	—	—	<0.5
E-RFC	1.7 ± 1.6	3.3 ± 0.8	95.8 ± 2.8	6.3 ± 4.3	8.2 ± 4.1
Non-E-RFC	—	—	1.2 ± 0.6	37.6 ± 15.6	33.7 ± 24.5
IgG-anti-IgG column-filtrated cells	<0.1	<0.1	96.2 ± 3.1	—	—

[a] Mean value (%) from six experiments ± standard deviation.

[b] Surface Ig determined by radioautography using a mixture of ^{125}I-F(ab')$_2$ antihuman γ-, μ-, δ- and a-chain.

CMC after rosetting (unpublished results). Furthermore, EAC-rosette centrifugation had to be performed at 37°C, since SRBC were used as antibody-coated indicator cells. Therefore, sedimentation at room temperature resulted in E-rosette formation by T lymphocytes carrying high-affinity SRBC receptors.

The T-cell nature of the cytolytic effector cells described here and by West *et al* (7) rests solely on the demonstration of the E-rosette marker. In order to characterize further these cells as members of the T-cell lineage, they were treated with a specific rabbit-antihuman thymocyte globulin (ATCG) and complement. These antibodies lysed 100% of the cells of T-lymphoblastoid cell lines but less than 10% of CLL lymphocytes or B-lymphoblastoid cell lines (12). When NAL were treated with ATCG and complement, no significant loss of CMC was observed. The almost identical dose response curves in Fig. 4 suggest that the cells

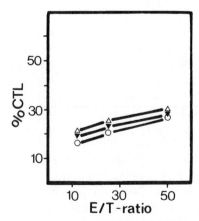

FIG. 3. Spontaneous cytolytic activity of NAL (○) following depletion of 19S-EAC3b (▾) or 19S-EAC3d (△) binding lymphocytes by rosette sedimentation. Each point represents the mean value of three experiments. Target cell Mel-Ei-78.

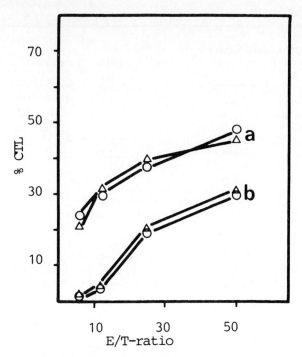

FIG. 4. Spontaneous cytolytic activity of NAL before (○) and after (△) treatment with ATCG and complement. The results of two experiments (a,b) are shown.

not lysed by ATCG are as cytotoxic as untreated NAL. Under the assumption that the ATCG-resistant cells are non-E-RFC, this result is at variance with the finding in Fig. 1, which shows that CMC of non-E-RFC is higher than that of E-RFC or even NAL. To resolve this discrepancy, purified E-RFC were treated with ATCG and complement as described. It was found that 8-15% of E-RFC were resistant to lysis. This proportion of ATCG-resistant E-RFC could be reduced neither by prolonged exposure to ATCG and complement, nor by higher ATCG concentration, nor by use of complement from other species. To avoid the influence of dead cells on the cytolytic system, they were separated from ATCG-resistant E-RFC by density centrifugation. By this additional experimental step, however, the recovery of ATCG-resistant E-RFC was lowered to about 1% of NAL (see Table 2). The presence of SRBC receptors on these ATCG-resistant cells was demonstrated by repeated rosetting experiments. For this purpose, it was necessary to preincubate the cells for at least 26 hr at 37°C, with or without prior trypsinization.

Despite the resistance of these particular E-rosette-forming cells to lysis by ATCG and complement, the presence of T-cell antigens on their surface is to be inferred from the finding that E-rosette formation could be completely inhibited by ATCG without complement. Such inhibition experiements are depicted in

TABLE 2. Cytolytic Activity of ATCG-Resistant Cells [a]

Experiment No.	Lymphocyte preparation	% Recovery cells from NAL	% E-RFC in (re-) rosetting experiments	Number of lytic units per 10^7 cells
1	NAL	100.0	76	110
	E-RFC (total)	21.4	95	34
	E-RFC (ATCG/C-resistant)	0.8	7^b $(97)^c$	29
	Non-E-RFC (total)	6.7	3	50
	Non-E-RFC (ATCG/C-resistant)	5.3	0	19
2	NAL	100.0	72	30
	E-RFC (total)	50.0	98	12
	E-RFC (ATCG/C-resistant)	1.2	0^b $(97)^c$	15
	Non-E-RFC	10.0	0	32
	Non-E-RFC (ATCG/C-resistant)	5.2	0	27

[a] Cell fractions were prepared as described in "Materials and Methods." One ly. unit was defined as the number of lymphocytes required to achieve 25% target cell lysis. Target cell: Mel-Ei-78. Percent revocery of cells was determined from the number of viable cells recovered in each fraction compared to the total number of original cells before fractionation.

[b] % E-RFC determined immediately after ATCG/C treatment.

[c] % E-RFC after incubation of ATCG/C resistant cells in RPMI 1640 medium supplemented with 10% FCS for 16 hr at 37°C.

Fig. 5. They show clearly that the antibody binds to *all* E-RFC. This effect of ATCG could be removed by absorption with human thymocytes. Since the lysis of lymphocytes was performed in antibody excess, the lytic supernate was also found to inhibit E-rosette formation to an extent corresponding to the residual antibody activity.

When the ATCG-resistant subpopulation of E-RFC was tested in the cytolytic assay, a CMC comparable to that of untreated E-RFC was found. This has been shown in six independent experiments. In Table 2 the data of two representative experiments are given. These results and the above findings suggest that the relevant CMC activity of T lymphocytes resides in a subpopulation that forms E rosettes and bears FcR and T-cell antigen(s) but is resistant to lysis by ATCG plus C. Although this subpopulation of CMC effector cells was enriched more than ten times in the ATCG-resistant E-RFC population, a corresponding increase in CMC activity could not be observed. In order to see whether this lack of enrichment was caused by inhibitory factors generated during cell lysis, NAL were incubated in heat-inactivated ATCG/complement lytic supernates under conditions of complement lysis, washed, and tested as effector cells in CMC and

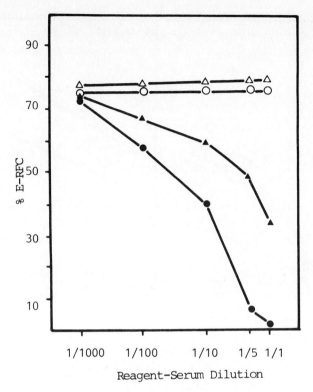

FIG. 5.   Inhibition of E-rosette formation by ATCG. Inhibition assays were done in duplicate. Prior to E-rosette formation, 1 x 10^6 lymphocytes were incubated for 90 min at room temperature in 100 μl of either FCS (o), ATCG (•) or the ATCG/C lytic supernate (▲) of 1 x 10^8 lymphocytes. All reagents were heat inactivated at 56°C for 60 min. ATCG/C-lytic supernate was absorbed with thymocytes at 37°C and 4°C for 30 min. (△).

ADCC. As shown in Fig. 6, CMC activity was found to be considerably reduced, whereas ADCC activity was completely abolished. Treatment with ATCG alone did not affect CMC by NAL although their ADCC activity was markedly inhibited.

From the data presented thus far one would expect that ATCG-resistant E-RFC contain a high proportion of FcR-positive cells that bind aggregated IgG or can be stained with anti-Ig reagents when armed with antibody. For detection of such cells, autoradiographic studies were performed allowing the simultaneous demonstration of E-rosette formation and binding of either ^{125}I-aggregated IgG or ^{125}I-F(ab')$_2$-antihuman-Ig antibodies specific for γ or μ chains. In these double marker studies, however, no FcR-bearing or Ig-positive E-RFC could be demonstrated in the ATCG-resistant cell population. In control experiments the same antibody preparation stained more than 90% of the non-E-RFC.

Although these data do not support the concept of *in vivo* armed T-marker bearing effector cells, E-RFC were found in a number of melanoma patients that reacted with ^{125}I-F(ab')$_2$-anti-γ-chain-reagents. With ^{125}I-F(ab')$_2$-anti-μ-chain

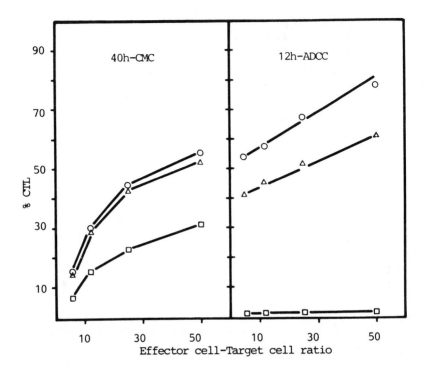

FIG. 6. Effect of ATCG or ATCG/C-lytic supernate on CMC and ADCC. Before the cyto-toxic test NAL (1×10^7/ml) were pretreated with either ATCG (1 : 4) ($\triangle$) or the ATCG/C-lytic supernate ($\square$) of 1×10^8 NAL. The cells were washed twice. All reagents were heat in-activated at 56°C for 1 hr. ADCC was measured by direct addition of a homologous anti-melanoma antiserum to the cytolytic system.

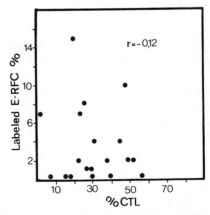

FIG. 7. Correlation between cytolytic activity of NAL and the percentage of E-RFC labeled with ^{125}I-Goat F(ab$'$)$_2$ antihuman γ chain in radioautography.

TABLE 3. Correlation Between the Percentage of γ-Bearing E-RFC in the Peripheral Blood of 20 Melanoma Patients and the Content of Antimelanoma Antibodies in Patient's Sera Detected by Indirect Immunofluorescence

Melanoma patients	γ-Positive E-RFC (% ± SD)
with antimelanoma antibodies	6.3 ± 5.4
	$p = <0.05$
Without antimelanoma antibodies	1.6 ± 1.6

antibodies or with ^{125}I-F(ab')$_2$ -preparations of normal IgG, no labeled E-RFC were detected. The double marker lymphocytes appeared to be larger than the nonlabeled E-RFC. The percentage of such cells did not correlate with the level of patients' CMC-activity (Fig. 7). In additional experiments, however, a strong correlation between the percentage of double marker cells in patients' blood and the presence of antimelanoma antibodies in patients' serum was observed as depicted in Table 3.

In another approach, monovalent fragments of immunoabsorbent purified anti-Ig antibodies were used to inhibit CMC reactions. The results of this extensive study are reported elsewhere in detail (14). The data of one experiment are given in Table 4 illustrating that Fab-antihuman IgG preparations failed to inhibit CMC in a constant and dose-dependent manner.

DISCUSSION

In previous experiments we showed that FcR-positive E-RFC can function as effector cells in an ADCC system consisting of human IgG antibodies and melanoma tumor cells. From the data of that study, it was tempting to speculate that

TABLE 4. Effect of Immunoadsorbent-Purified Goat Fab-Antihuman IgG Preparations on CMC[a]

Lymphocyte donors	CMC-Activity (% CTL)	% Inhibition of CMC by Fab-antihuman IgG		
		0.02	1.0	0.1 mg/ml
I.D. (melanoma)	30	0	0	6
A.S. (melanoma)	34	0	0	0
G.J. (melanoma)	22	54	56	64
E.G. (melanoma)	21	39	52	43
J.S. (healthy)	35	0	0	0

[a] CMC activity of NAL was measured in a 40 hr ^{3}H-proline release test at an effector cell-target cell ratio of 50 : 1. Target cell: Mel-Ei-78.

T cells armed with antibody mediate spontaneous cytotoxicity (6). This hypothesis is partly confirmed by the present study to the extent that a portion of
the total spontaneous cytotoxicity is effected by FcR-bearing lymphocytes binding SRBC under optimal conditions. In contrast to the major proportion of human
T lymphocytes, this subpopulation is resistant to lysis by a heterologous antithymus cell globulin and complement. The presence of relevant T-cell antigens
was demonstrated by the complete inhibition of E-rosette formation with ATCG.
One possible explanation for their survival in the cytotoxic test may be a low
density of T-cell antigens on the cell membrane. This line of reasoning is supported by recent findings that show that natural cytotoxicity in the mouse is
mediated by cells expressing low-density Θ antigen [for review see ref. (4)]. The
T-cell nature of these effector cells is also supported by the data of West *et al.*
(7) showing that FcR-positive cells that bind SRBC with low affinity mediate
human natural cytotoxicity.

Complement receptor-bearing lymphocytes seem not to be of major importance in our cytotoxic system. The studies reporting the presence of complement
receptors on human effector cells rest on depletion experiments using EAC rosettes (3). From our experience it is possible that such effector cells are removed
via their Fc receptors when the IgM antibodies used for sensitization of SRBC
are contaminated with minute amount of IgG antibodies. One should bear in
mind, however, that in our system not only the described E-RFC subpopulation
but also non-E-RFC are effective in CMC. This observation indicates that a spectrum of different cell types is involved in lytic reactions, the actual activity depending on the experimental system investigated. It is difficult to assess the relative cytolytic potential of each of the various cytotoxic cell types, however, since
the separation techniques themselves alter the cytolytic capacity as well as the
expression of surface markers of the isolated cell population. Thus, prolonged incubation of NAL at 37°C (1, 4, 15) as well as treatment of E-RFC with ammo-
nium chloride (7) or hypotonic solutions (15) to remove SRBC contamination
result in CMC depression. The fact that the latter procedure leaves intact the
capacity of cells to cooperate with antibody in ADCC suggests that by hypotonic treatment a recognition structure may be removed from the cell surface
that is probably arming antibody. Similar considerations have been reported by
Akira *et al.* (8). Furthermore, as already shown, treatment of NAL with ATCG/
complement lytic supernates produced a marked decrease of CMC as well as
ADCC, most likely caused by inhibition of FcR activity by immune complexes
liberated during cell lysis. Treatment of lymphocytes with ATCG alone, however,
led exclusively to ADCC inhibition. This may be due to competition of the
antilymphocyte and the antitumor antibodies for FcR sites on ADCC effector
cells, a situation that has been described in the cytotoxic inhibition assay by
Halloran *et al.* (16). Such experimentally induced inhibition phenomena of CMC
activity may explain why the expected enrichment of CMC activity in the ATCG-
resistant E-RFC population was not seen. Furthermore, in view of the particularly cumbersome handling of the E-rosette forming cells, it is not surprising
that several workers (2, 3) found human CMC competence restricted to the

non-E-RFC fractions. As mentioned before, not only the cytolytic potential of cells but also surface markers are altered by cell separation proceudres. Thus, it is shown here that the E-rosette forming capacity is impaired after treatment with ATCG and complement on the surviving cell population. Furthermore, the autoradiographic demonstration of FcR is obviated by immune complexes generated during this complement-induced lytic process. The failure to detect FcR- and/or S-Ig-positive cells in the ATCG-resistant E-RFC fraction may also be explained by the findings of Moretta *et al.* (17) who showed an extreme *in vitro* lability of this Ig-binding membrane structure.

The modulation of FcR on T lymphocytes by immune complexes may also account for the difficulties in demonstrating an *in vivo* arming of effector cells. Therefore, experiments concerning the existence of an *in vivo* arming mechanism have to be interpreted with caution, particularly the inhibition experiments with Fab-anti-Ig reagents.

Acknowledgments

We thank M. Babbel, B. Ermler, and F. Frank for excellent technical assistance and C. Krauss for help with the manuscript.

References

1. Bonnard, G. D., and West, W. H. *In* "Immunodiagnosis of Cancer" (R. B. Herberman and K. R. McIntire, eds.). Marcel Dekker, New York.
2. Peter, H. H., Pavie-Fischer, J., Fridman, W. H., Aubert, C., Cesarini, J., Roubin, R., and Kourilsky, F. M. *J. Immunol.* **115**, 539, 1975.
3. Pross, H. F., and Jondal, M. *Clin. Exp. Immunol.* **21**, 226, 1975.
4. Herberman, R. B., and Holden, H. T. *Adv. Cancer Res.* **27**, 305, 1978.
5. Kay, H. D., Bonnard, G. D., West, W. H., and Herberman, R. B. *J. Immunol.* **118**, 2058, 1977.
6. Saal, J. G., Rieber, E. P., Hadam, M., and Riethmuller, G. *Nature* **265**, 158, 1977.
7. West, W. H., Cannon, G. B., Kay, H. D., Bonnard, G. D., and Herberman, R. B. *J. Immunol.* **118**, 355, 1977.
8. Akira, D. and Takasugi, M. *Int. J. Cancer* **19**, 747, 1977.
9. Greenberg, S. B., Criswell, B. S., Six, H. R., and Comes, R. B. *J. Immunol.* **119**, 2100, 1977.
10. Imir, T., Saksela, E., and Makela, O. *J. Immunol.* **117**, 1938, 1977.
11. Saal, J. G., Rieber, E. P., and Riethmuller, G. *Scand. J. Immunol.* **5**, 455, 1976.
12. Rodt, H., Thierfelder, S., Thiel, E., Gotze, D., Netzel, B., Huhn, D., and Eulitz, M. *Immunogenetics* **2**, 411, 1975.
13. Dierich, M. P., and Bokisch, V. A. *J. Immunol.* **118**, 2145, 1977.
14. Hadam, M., and Saal, J. G., in preparation.
15. Saal, J. G. *et al.*, in preparation.
16. Halloran, P., Schirmacher, V., and Festenstein, H. *J. Exp. Med.* **140**, 1348, 1974.
17. Moretta, L., Ferrarini, M., and Cooper, M. D. *Contemp. Top. Immunobiol.*, **8**, 19, 1978.

RELATION OF EFFECTOR MECHANISMS TO TARGET CELL ORIGIN

M. Troye, G. R. Pape, M. Vilien, and P. Perlmann

Cell-mediated immune reactions *in vitro* to a variety of human and animal tumors have been studied extensively during recent years (1). The most commonly used *in vitro* approaches have been to assess lymphocyte cytotoxicity against a multiplicity of cultured cell lines. However, the issue of specificity of these reactions and of their biological significance remain controversial (2). The cytotoxic effector cells active in many of these systems are not well characterized. The aim of these studies was to establish the occurrence of a *disease-related* lymphocyte cytotoxicity (3) in patients with transitional cell-carcinoma (TCC) of the urinary bladder *in both allogeneic and autochtonous* lymphocyte-target cell combinations and to characterize the effector cell(s) operative in these systems.

In previous studies, highly purified peripheral lymphocytes (PBL) from four different groups of donors, each comprising approximately 20-30 individuals, were tested in a ^{51}Cr-release assay against a small panel of allogeneic target cells from established cell lines (4). The different groups of donors were: (a) untreated TCC-bladder patients all with tumors of clinical stage T1-T3; (b) TCC-bladder patients treated with local radiotherapy 1-12 yr before testing; (c) age- and sex-matched clinical controls, mostly patients with cancer of the prostate, also untreated; and (d) healthy donors. The target cell panel consisted of five different cell lines, two of TCC-bladder origin, distinct from each other in karyotypes and in HLA antigens (unpublished observations), one from colon carcinoma and one from malignant melanoma. Percentage cytotoxicity was expressed by subtracting the percentage background cytotoxicity in the lymphocyte-free medium controls from that in the lymphocyte-containing experimental samples.

57

The lymphocytes of the TCC-bladder patient group had stronger mean cytotoxicity against the allogeneic tumor-target cells of TCC-bladder origin than the
lymphocytes of the two control groups. The differences were statistically significant (4). Thus, TCC-bladder patients as a *group* show *disease-related* cytotoxicity. However, individual donors from all groups were frequently cytotoxic for
target cells of different types. This individual cytotoxicity was *selective*, i.e., it
was directed against one or a few targets. *Nonselective* (i.e., generalized, "nonspecific") cytotoxicity (3) was rare. Our results thus suggest that this lymphocyte cytotoxicity reflects a reaction of both patients and healthy donors to a
variety of complex determinants, differently expressed on various targets. In
principle these determinants would be of at least three types, namely, (a) tumor-
associated, (b) tissue-specific, and (c) "species specific" (human). These operationally defined antigens would be expected to consist of several specificities;
cross reactions would be common. These issues are presently under investigation.

Effector cell studies in these systems were performed in allogeneic combinations against the aforementioned target cell panel by fractionating highly purified PBL on different columns separating lymphocytes according to some of
their known surface markers (5). Removal of SIg^+ (B cells) did not significantly
affect cytotoxicity, but when lymphocytes with Fc receptors for IgG (FcR^+)
were removed, cytotoxicity was abolished. Furthermore, cytotoxicity was in virtually all cases inhibited to 60-100% when Fab fragments of rabbit antihuman
immunoglobulin (a-HuIg) were added to the incubation mixture. These results
suggest the involvement of effector lymphocytes operating via antibodies (K
cells) (6). This was the case when lymphocytes of TCC-bladder patients, clinical
controls and healthy donors were tested against allogeneic targets (7). However,
when lymphocytes from the TCC-bladder patients were tested in both allogeneic
and autochthonous combinations, the pattern in the autochthonous combination was different. Neither removal of FcR^+ or SIg^+ lymphocytes affected the
cytotoxicity against the autochthonous tumor target cells. However, when the
same TCC-bladder target cells were tested in allogeneic combinations, the results
followed the patterns described, i.e., removal of FcR^+ lymphocytes abolished
cytotoxicity. Furthermore, when lymphocytes from these two patients after depletion of FcR^+ cells were tested in allogeneic combinations, no cytotoxicity was
seen. Thus, while a significant part of the cytotoxicity against allogeneic target
cells appears to be an antibody-dependent K-cell mediated reaction, the effector
mechanism in the autochthonous combination seems to be different. The lymphocytes operative in the autochthonous reactions could be antibody-independent
cytolytic T cells (CTL), displaying a syngeneically restricted cytotoxicity as seen
in animal tumor systems (8). Alternatively, this cytotoxicity could be due to
antibody-independent NK cells (9) or N cells (10). Other mechanisms should
also be considered (11-13). To further investigate the effector cells involved in
autochthonous combinations, additional target cells and lymphocyte donors
have now been investigated. The cell lines used in this study are listed in Table
1. These lines were divided into four groups: (a) three cell lines of TCC-bladder
origin established several years ago; (b) one cell line from normal bladder epi-

TABLE 1. Characteristics of Target Cells

Group[a]	Cell Line	Origin	Type[a]	Donor Available
1	T-24	TCC[b]	"T-24"	−
	Hu 456	TCC	"T-24"	−
	Hu 549	TCC	"T-24"	+
2	Hu 609	NHB[b]	e	+
3	Hu 961	TCC	e	+
	Hu 1244	TCC	e	+
	Hu 1264	TCC	e/f	+
	Hu 1125	TCC	e/f	+
	Hu 1210	TCC	(e)/f	E
4	Mel-1	Melanoma	e	−

[a] Type: "T-24" = morphologically similar to T-24; e = epitheloid; f = fibroblastlike. Group: 1,2, and 4-established cell lines; 3 = short-term cultures, 10-20 passages (see text).
[b] TCC, transitional cell carcinoma of the urinary bladder; NHB, normal human bladder epithelium.

thelium, from a patient with hypernephroma; (c) five TCC-bladder cell lines established in culture half a year before testing. These cells were all in passage number 10-20 and are referred to as *short-term* cultures. However, the cells in this group are also distinct from primary cultures as discussed by Baldwin and Embleton (2). The short-term cultures in group (c) were all growing at an extremely slow rate and sometimes gave high-spontaneous release of isotope in the ^{51}Cr-release assay. (Experiments in which the target cells released $\geq 25\%$ ^{51}Cr spontaneously were excluded.) (d) The fourth group is an established cell-line of melanoma origin . All cell lines except T24 (14) and Mel-1 (15) were established by one of us (M.V.). Cell lines that were similar to T24 as judged morphologically by light microscopy were denoted "T24-like." Some of the short-term cultures appeared to be mixtures of epithelial and fibroblastlike cells. All experiments were set up in both autochthonous and allogeneic effector : target cell combinations. Because of the slow growth rate of the short-term cell lines, fewer experiments were done with these cells than with those from the established lines and in these instances the results could not be evaluated statistically.

The results obtained with the established cell lines for three different groups of donors are presented as scatter diagrams in Fig. 1. The lymphocytes from TCC-bladder patients had a significantly elevated reactivity against T24, when compared with the other donors (unpaired t-test, $p \geq 0.05$). Their cytotoxicity for the other TCC-bladder line was also elevated but this was not statistically significant, probably due to the smaller number of donors tested against this cell line. No differences between the other donor groups with regard to cytotoxicity for the other two targets could be seen. When the data were assessed by a paired t-test, no significant differences in reactivity for the TCC-bladder group was apparent between T24 and Hu 456, whereas comparison of T24 with Hu 609 and

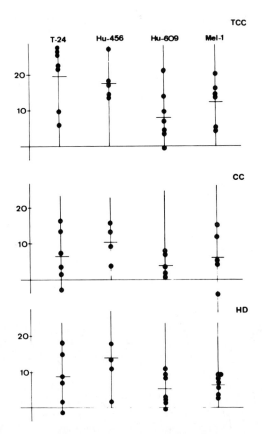

FIG.1. Cytoxocity of lymphocytes from three different groups of donors to four different target cells. Each dot represents the cytotoxicity of one individual donor's lymphocytes. Horizontal lines: mean cytotoxicity. Abscissa: % ^{51}Cr-release from experimental samples corrected by subtracting of % release in lymphocyte free medium controls. See also (4). TCC = TCC-bladder patients; CC = clinical controls; HD = healthy donors.

T24 with Mel-1 did show such differences ($p < 0.01$ and $p < 0.05$, respectively). TCC-bladder patients whose lymphocytes had a high reactivity against T24 had this also against Hu 456. This pattern was not found within the two control groups. By comparing the data with a paired t-test, it was also concluded that T24 was more susceptible to lysis than the other cell lines. Had this been the case, a significant killing of T24 in comparison to the other cell lines should have been found among the two control groups.

The results of experiments in which lymphocytes of four different donors were tested in both autochthonous and allogeneic combinations are summarized in Table 2. In three of four cases the cytotoxicity against T24 and the corresponding autochthonous target cells was not correlated, suggesting involvement of different determinants. In two of four cases there was no reactivity in the autochthonous combinations. When SIg+ lymphocytes or FcR+ lymphocytes

TABLE 2. Cytotoxicity of Patients' Lymphocytes to Allogeneic
or Autologous Bladder Tumor Target Cells[a]

	U				B⁻				FcR⁻			
	ALLO		AU		ALLO		AU		ALLO		AU	
Donor	−	+	−	+	−	+	−	+	−	+	−	+
1244	27.7	1.8	−2.1	2.8	22.1	0.4	0.7	−	6.3	−3.6	2.7	4.0
961	21.6	5.6	26.8	17.0	13.6	0.7	27.0	16.1	9.3	2.5	22.2	19.7
549	19.0	7.9	2.4	1.1	10.5	0.4	−2.1	−	6.0	9.5	0.8	−
1125	9.8	8.2	19.6	16.2	5.2	8.6	20.5	18.8	4.5	11.5	20.2	18.5
Mean	18.2	5.9	11.7	9.2	12.9	2.5	10.2	17.5	6.5	5.0	11.5	14.1

[a] TCC patients. U: unfractionated lymphocytes; B⁻: depleted of B cells; FcR⁻: depleted of FcR⁺ lymphocytes, ALLO; AU: target cells of allogeneic (T24) or autologous origin, respectively. −: No inhibitor present; +: Immunoadsorbent purified Fab-anti-HuIgG added to incubation mixture (100 μg/ml). For details see (6). The numbers are % ^{51}Cr-release (corrected) after 18 hr of incubation at a lymphocyte : target cell ratio of 60:1. For details see (4) and (6).

were removed by column fractionation (5, 7), the cytotoxicity pattern against the allogeneic target cells (T24) was as usual, i.e., cytotoxicity was partially or not abolished by removal of SIg⁺ cells, whereas removal of FcR⁺ cells reduced cytotoxicity. Furthermore, in three out of four cases allogeneic cytotoxicity was inhibited by Fab fragments of rabbit a-HuIg, again suggesting a K-cell mechanism (6). In contrast, in the two positive autochthonous combinations, neither removal of SIg⁺ cells nor of FcR⁺ cells affected cytotoxicity. Moreover, this autochthonous cytotoxicity was not significantly reduced when Fab fragments of rabbit a-HuIg were added to the incubation mixtures. These results are in agreement with our earlier findings (7) suggesting an antibody-independent mechanism operative in autochthonous combinations. However, it is noteworthy that the sensitivity of target cells from short-term cultures was more variable with reactive allogeneic lymphocytes from clinical controls or healthy donors. Thus, in three of six such preliminary experiments, removal of FcR⁺ lymphocytes did not significantly reduce cytotoxicity.

CONCLUDING REMARKS

Cytotoxicity to *allogeneic target cells from established cell lines* is frequently, but not invariably due to effector lymphocytes of the K-type lysing target cells through an immunoglobulin-dependent mechanism. This holds for both the disease-related reactions of cells from TCC-bladder patients and for the selective cytotoxicity of lymphocytes from clinical controls or healthy donors. In four of six cases, effector cells from TCC-bladder patients acting on *autologous TCC-bladder target cells* were antibody-independent cells, i.e., they had no Fc receptors for IgG and were not inhibited by Fab fragments of rabbit antibodies to human IgG. In two of six cases no autochthonous cytotoxicity was discussed. Since the autologous reactions in three out of four positive cases were stronger than the allogeneic reactions, this may reflect a specific CTL-mediated cytotoxicity, but direct proof for CTL participation is lacking and other mechanisms should be considered. This judgment is also warranted by the finding that target cells from short-term cultures of TCC-bladder cells were in some instances lysed by allogeneic FcR⁻lymphocytes from clinical controls and healthy donors. This exceptional behavior could be due to some heterogeneity at the target cell level (see Table 1) or to the particular lymphocyte-target cell combinations in which cytotoxicity was caused by FcR⁻effector cells. It is also possible that in some of these instances cytotoxicity was due to the generation of FcR⁺ effector cells during the assay (11), or, conversely it could reflect yet a different type of mechanism such as described by Trinchieri et al (13).

Acknowledgment

This work was supported by grant no. 365-B77-09XA from the Swedish Cancer Society.

References

1. Hellstrom, K. E., and Hellstrom, I. Adv. Cancer Res. 12, 167, 1969.
2. Baldwin, R. W., and Embleton, M. J. Int. Rev. Exp. Pathol. 17, 49, 1979.
3. Bean, M., Bloom, B., Herbermann, R., Old, L., Oettgen, H., Klein, G., and Terry, W. Cancer Res. 35, 2902, 1975.
4. Troye, M., Perlmann, P., Larsson, A., Blomgren, H., and Johansson, B. Int. J. Cancer. 20, 188, 1977.
5. Pape, G. R., Troye, M., and Perlmann, P. J. Immunol. 118, 1919, 1977.
6. Troye, M., Perlmann, P., Pape, G. R., Spiegelberg, H. L., Naslund, I., and Gidlof, A. J. Immunol. 119, 1061, 1977.
7. Pape, G. R., Troye, M., and Perlmann, P. J. Immunol. 118, 1925, 1977.
8. Cerottini, J. Ch., and Brunner, T. K. Adv. Immunol. 18, 67, 1974.
9. Kiessling, R. E., Klein, E., Wigzell, H. Eur. J. Immunol. 5, 112, 1975.
10. Kiuchi, M., and Takasugi, M. J. Natl. Cancer Inst. 56, 575, 1976.
11. Saal, J. G., Rieber, E. P., Hadam, M., and Riethmuller, G. Nature 265, 158, 1977.
12. Peter, H. H., Eife, R. F., and Kalden, J. R. J. Immunol. 116, 342, 1976.
13. Trinchieri, G., Santoli, D., and Koprowski, H. 1977.
14. Bubenik, J., Baresova, M., Viklichy, V., Jakoubkova, J., Sainerova, H., and Donner, J. Int. J. Cancer. 11, 765, 1973.
15. Unsgaard, B., and O'Toole, C. Brit. J. Cancer. 31, 301, 1975.

MODELS FOR MECHANISMS OF NATURAL CYTOTOXICITY

Guy D. Bonnard, H. David Kay, Ronald B. Herberman,
John R. Ortaldo, Julie Djeu, Kathy J. Pfiffner, and
J. Ronald Oehler

The concept of natural cell-mediated cytotoxicity in humans derived from observations that peripheral blood mononuclear cells of normal healthy donors, and cancer patients as well, mediated direct cytotoxic reactions against a variety of neoplastic and normal cell lines. The presence of these natural cytotoxic cells [reviews in (1-3)] has complicated the study of specific, disease-related direct cytotoxic reactions. On the other hand, these natural killer (NK) cells are themselves of considerable interest, and they have been extensively studied in both murine and human experimental systems (1-6). Until recently, NK activity was thought to be mediated by lymphocytes devoid of detectable cell surface markers, which therefore appeared distinct from cytotoxic T lymphocytes and from antibody-dependent cell-mediated cytotoxicity (ADCC) (4, 5). More recently, studies of human (7-9) and mouse NK cells (3, 10, 11) have revealed that NK cells do bear an Fc receptor (FcR) and are in fact not separable from K lymphocytes; thus they may well represent the very same cells. Both types of activities can be adsorbed via FcR to immobilized antigen-antibody complexes, and this is reversed by the protein A of *Staphylococcus aureus*, (8, 11). NK and K cell activities were both found in the same lymphoid organs, namely, in peripheral blood and spleen, but not in tonsil, lymph node, or thymus (7, 9, 10). We also found that FcR-positive NK and K cell activities both persisted *in vitro* for 4-7 days, and both were also generated from nonactive precursors in lymphocytes cultured in media containing fetal bovine serum (12). In addition, both human NK and K cells were

63

shown to have low-affinity receptors for SRBC, forming rosettes under optimal conditions (4°C and high SRBC to lymphocyte ratios), but not under suboptimal conditions (e.g., at 29°C) (7, 9). Similarly, NK activity in the mouse was reduced after treatment with anti-Thy 1.2 sera plus complement, but usually only upon repeated treatment with high concentrations of antiserum (13). Thus, the natural cytotoxic effector cells in both humans and mice have been found to have analogous, low-density or low-affinity cell surface markers related to T cells. A more detailed consideration of these interspecies analogies has been reported elsewhere (3, 10).

Several investigators, particularly those working with human NK cells, have suggested that the mechanism of NK and ADCC could be the same. They have explored the possibility that NK could be mediated by arming antibodies attached to K cells. There is some evidence for involvement of immunoglobulin in NK, but conclusive evidence for the arming hypothesis is lacking. In addition, the actual mechanism of killing in ADCC as well as in NK is not known and deserves considerably more attention. Accordingly, we have developed several NK models, some with a partial role for Ig, in which emphasis is placed on underlying mechanisms of lysis. In these models, several different roles for Ig in addition to arming natural antibodies have been considered. A model, emphasizing similarities between results of studies on NK and on lymphotoxins, is presented elsewhere (14). In this particular model, there is no role for Ig whatsoever; as most experiments with anti-Ig reagents were negative, the specificity of NK could be explained without implicating Ig.

PROBLEMS WITH A MODEL FOR NK IN WHICH K CELLS ARE ARMED WITH NATURAL ANTIBODIES VIA THE CLASSICAL FcR

At this conference evidence is given for a role of Ig, arguing that such Ig may represent specific antibodies capable of inducing ADCC against coated target cells. Perlmann and associates have studied spontaneous cytotoxicity in an 18-hr ^{51}Cr-release assay against bladder carcinoma-derived cell lines. They compared the effector cells with K cells, and found them in the same subpopulation of FcR-positive lymphocytes (15-17). Subsequently, Ig was implicated since NK reactions were blocked by monovalent Fab fragments of rabbit antibody raised against human Ig or its F(ab')$_2$ and Fc fragments (18). Some reactions were not blocked by Fab anti-Ig, but this was thought to reflect specific immune T-cell cytotoxicity (18, 19). Investigators in Riethmüller's group used a long-term ^{3}H-proline release assay with melanoma-derived cell lines and were successful in partially blocking NK with Fab anti-Ig reagents (20). It is noteworthy that not all reagents of this kind were able to block NK activity. Mouse NK could not be inhibited by a potent and anti-Ig reagent that was quite effective in blocking ADCC (4) or by a potent Fab anti-F (ab')$_2$ reagent kindly provided by Dr. T. Chused. We have also been consistently unsuccessful in our K-562 system with either the $F(ab')_2$ or the Fab fragments of a highly specific commercially available goat anti-human $F(ab')_2$, known to interfere strongly with a potent, multivalent human anti-HLA serum, and which itself has no detectable anti-Fc activity (Kay et al., in preparation). This negative result does not seem attributable to the fact that we use K-562, a myeloid cell line, as target for NK, since Pape et al. will report at this conference an impressive blockade of NK against K-562 with other anti-Ig reagents (21).

The origin of the Ig detected with anti-Ig reagents is uncertain (15-17). Some time ago, we provided data obtained with the protein A of S. aureus that is informative in this regard. Protein A, when added for the 4-hr duration of our cytotoxicity assays, reduced ADCC, but it did not reduce NK (8). This failure of protein A to block NK could be explained by the presence of antigen-specific receptors on NK cells that do not require Ig (8). Alternatively, if NK cells utilize IgG antibodies via their FcR, they might pick them up prior to the assay, most likely in vivo, and these IgG molecules might then be unavailable for interaction with protein A (8). This is compatible with arming of NK cells in vivo with cytophilic antibodies attached to the FcR (8, 22-26). However, the evidence that in vivo bound cytophilic antibodies can induce ADCC is not clear, and at best indirect (25, 26). NK reactivity was shown by Takasugi's group to be specific and was documented in cold target cell inhibition studies as well as in direct cytotoxicity assays. When lymphocytes were trypsinized, they lost their NK reactivity, a finding interpreted as the consequence of removal of the cytophilic antibodies. Subsequent incubation of these pretreated cells with different human sera frequently, but not invariably, restored the NK activity. This effect was not obtained with sera that had been preadsorbed with the target cells. In addition, specific arming was also achieved in vitro, using well-defined anti-HLA sera (27). This is consistent with other reports on in vitro arming immune complexes or aggregated immunoglobulin (28, 29), but it does not directly bear on the question of in vivo arming by native Ig and natural antibodies.

We have studied in some detail the possible role of Ig in both the human K-562 and the mouse NK systems. Like others, we have observed that both human NK and ADCC are inhibited by sEA (sheep erythrocyte-antibody complex) or preparations of Ig (7, 30-33), and by cEA (chicken erythrocyte-antibody complex) (Table 1). However, there are no data to indicate that this mechanism of inhibition is by competition with, or displacement of, natural antibodies from FcR. In fact, we have accumulated data showing that interaction with FcR is not a generally effective means to inhibit NK. Several particulate and soluble immune complexes failed to inhibit NK, even though they did bind to the effector cells and were able to completely block ADCC (Table 1). Subsequently, in a large series of experiments, we sought to detect autologous lymphocyte-dependent antibodies in the sera of normal individuals, that would produce ADCC against K-562 when kept in the reaction mixture. In these instances, the presence of 10% autologous, fresh, nonheat-inactivated, normal human serum resulted in a moderate reduction in natural cytotoxicity, and we could not detect a boost of reactivity due to ADCC. In another series of assays, however, NK activity against

TABLE 1. Effects on Human NK or ADCC of Various Antisera or Immune Complexes[a]

Pretreatment of effector cells	NK	ADCC
E:Rabbit antisheep E complexes	Inhibited	Inhibited
E:Rabbit antichicken E complexes	Inhibited	Inhibited
Aggregated human IgG	Inhibited	Inhibited
Human anti-HLA serum (multivalent)	Unchanged	Inhibited
In the assay		
Human anti-HLA serum (multivalent)	Unchanged	Inhibited
Rabbit antihuman (Chang)	Unchanged	Inhibited
TNP:Rabbit anti-TNP complexes	Unchanged	Inhibited
Horse antilymphocyte serum	Unchanged	Inhibited

[a] Kay et al., in preparation.

K-562 was boosted on occasion when 50% allogeneic, fresh, nonheat-inactivated serum was used in the assay. Similar results were obtained in our mouse NK system with RL♂1 as target cells. When spleen cells from C57BL/6 animals were incubated with RL♂1 in the presence of high concentrations of mouse serum, the level of cytotoxicity observed was higher than in tests done in the absence of serum (Djeu et al., in preparation). These reactions may or may not reflect a postulated role of Ig in NK. However, we have not seen activity enhanced over control, if after preincubation, the excess serum was removed instead of kept in the assay cell mixture. Thus, we have essentially failed to demonstrate serum effects that could unequivocally be called arming. A typical experiment with human cells is given in Table 2.

TABLE 2. Attempts to Strip and Arm Effector Cells Active Against K-562[a]

| Treatment of effector cells | | Additives in assay | % ^{51}Cr Release |
First	Second		
None	—	—	45.0
	—	Serum	37.0
	Serum, 4°C	—	41.3
	Serum, 37°C	—	44.6
Held, 37°C, 1 hr	—	—	48.2
	—	Serum	37.0
	Serum, 4°C	—	42.0
	Serum, 37°C	—	44.0
Trypsinized	—	—	4.0
	—	Serum	5.0
	Serum, 4°C	—	2.0
	Serum, 37°C	—	5.0

[a] Kay et al., in preparation.

Other experiments on the *in vitro* generation of NK and K cells may help to exclude a role for Ig in NK. It has been possible to isolate FcR-negative lymphocytes on immobilized antigen-antibody complexes and such populations are devoid of NK and K cell activity (8). By culturing these cells with 10% fetal bovine serum for 7 days, high levels of NK and ADCC activities were recovered; such activity was again due to FcR positive cells (12). Recently, we examined the cellular requirements for such *in vitro* generation of NK and K cells (Ortaldo *et al.*, in preparation). After depletion of FcR-positive cells, the following subpopulations fractionated into SRBC rosette forming cells (E+), surface membrane immunoglobulin positive cells (SIg+), and null cells (lacking E+, SIg+, or FcR+ markers). Unseparated FcR-depleted cells developed considerable activity after 7 days in culture with fetal bovine serum, whereas each of the subpopulations cultured alone developed little or no reactivity. Addition of SIg+ cells to other cell types had *no* effect. Mixtures of 80% E+ and 20% null cells resulted in optimal NK and ADCC. Experiments with mixtures of irradiated and unirradiated subpopulations of cells, indicated that the null cells were the precursors of both NK and K cells and the E+ cells acted as essential helper cells. So far, the null cell precursors did not appear to belong to a B-cell subset, nor to represent SIg-negative plasmacytes (34). Antibody production is definitely not a critical feature of this system, and K cells are generated as well as NK cells. In addition to rendering unlikely a role for Ig in NK, these experiments indicate a clear distinction between the cell surface characteristics of the human NK effector cells and their precursors. The distribution of human NK effector cells in various organs has been well studied (7), and now the way is open to study the organ distribution of their precursors. It is important to be aware of these differences between the NK effector cells, and their precursors, and to develop separate, appropriate protocols to study separately each of them.

Another way to generate NK and K cells has been the coculture of lymphocytes with tumor cells (35-38). A study of this phenomenon (39, this conference) suggested a major role for interferon in inducing NK; ADCC did not appear to be increased in parallel (39). We have also sought to characterize the postulated role of interferon in inducing NK in both rats and mice. In rats, the levels of cell-mediated cytotoxicity against tumor cells could be augmented by intraperitoneal injection of a variety of interferon inducers, including *Corynebacterium parvum*, lymphocytic choriomeningitis virus, Kilham rat virus, and poly I: C (40, 42). The specificity of the augmented cytotoxicity appeared to be the same as the specificity of NK as present in normal rat spleen. The cells mediating the augmented cellular cytotoxicity were small, nonadherent esterase-negative lymphocytes with Fc receptors, as are rat NK cells. The kinetics of the augmentation of NK activity by poly I:C and *C. parvum* were found to differ, with a shorter time course of augmented activity seen after inoculation of poly I:C. These results suggest interferon augments NK *in vivo*. More experiments along these lines were also made for a distinction between NK effector cells and their precursors (41). Treatment of rats with high doses of hydrocortisone, X-irradiation, or cyclophosphamide had a suppressive effect on natural cytotoxicity *in vivo*. However, when rats were given poly I:C after any of these agents, the levels of NK activity were similar to those in normal rats that had been given poly I:C alone. To explain these findings, we have postulated that a population of precursors, resistant to hydrocortisone, cyclophosphamide, and X-irradiation was induced by poly I:C to become NK effector cells.

Similar *in vivo* boosting of NK reactivity was observed in mice after inoculation of *C. parvum*, lymphocytic choriomeningitis virus, poly I:C, and pyran copolymer (Djeu *et al.*, in preparation). Again, the action of poly I:C was more rapid (1 day) than the action of either *C. parvum* or lymphocytic choriomeningitis virus (3 days). Crude preparations of mouse interferon and a preparation purified by affinity chromatography (42) caused detectable augmentation of NK as early as 3 hr after intraperitoneal inoculation. These preparations were also very active *in vitro*. When lymphocytes were preincubated for only 45 min with either crude or purified preparations of interferon, significantly higher levels of NK were observed. Similar *in vitro* induction was obtained with poly I:C. The results did not change if the lymphocytes were thoroughly washed between the preincubation and the cytotoxicity assay. So far, we have not determined whether interferon causes a boost in ADCC in parallel with NK, but this seems likely since *in vivo* inoculation of LCMV or *C. parvum* causes a boost in both ADCC and NK (10). These findings of boosting of NK with preparations that either contain or induce interferon further argue against a major role for Ig in NK. They raise the possibility, which could be tested, that some of the boosting effects seen with sera could be mediated by interferon, rather than by Ig and arming.

The model of arming by Ig does not take into account recent findings on the lability of attachment of cytophilic antibodies. Indeed several kinds of evidence support the contention that FcR-positive lymphocytes with cytophilic Ig (22-24) are indeed the cells that mediate ADCC and NK. Conditions exist, under which the K cells retain their cytophilic Ig and are therefore adsorbed on F(ab')$_2$ anti-F(ab')$_2$ columns (43, 44). In recent experiments, we were able to demonstrate selective adsorption of both K and NK activities on plastic surfaces coated with our F(ab')$_2$ anti-F(ab')$_2$ reagent (Kay *et al.*, in preparation). However, studies on cytophilic Ig (22-24) on depletion of cells on F(ab')$_2$ anti-F(ab')$_2$ columns (44), all emphasize that cytophilic Ig is labile and is usually lost after 30 min incubation at 37°C. After such treatment, cytophilic antibodies are usually no longer found on effector K and NK cells using F(ab')$_2$ anti-F(ab')$_2$ reagents (43) or immunoadsorbents (15, 16, 44). This contrasts sharply with the persistence of human NK activity after incubation for several hours, or after thorough washings at 37°C (Kay *et al.*, in preparation). Furthermore, there is evidence that the FcR itself is rapidly inactivated or shed following interactions with immune complexes (45). Whereas this could explain the cessation of ADCC reactions after a few hours (38, 46), there was no evidence that NK reactions were in any way affected by a similar mechanism, since they continued for longer periods (38). In summary, therefore, the findings in these studies are difficult to reconcile with a model for NK in which K cells are armed by natural antibodies via the classical FcR. We have therefore developed other models that seem to better fit the available data.

MODEL NO. 1: NATURAL ANTIBODIES ARM FcR2 ON FcR1-FcR2-POSITIVE CELLS

Because our data indicate that several types of immune complexes have different effects on NK and ADCC (Table 2), and because of the incompatible shedding data discussed, we have come to think of models for killing that involve alternative types of Ig binding to the effector cells. One possibility is that NK cells have two types of FcR, FcR1 for ADCC reactions, and FcR2 for cytophilic antibodies on the same cells. In this model NK could be mediated by cytophilic antibodies on FcR2. The concept of heterogeneity of the FcR is not new, going back to the studies of Fröland *et al.* (47), who were concerned with differences between FcR on B and K cells. Ig might bind to various types of FcR with different affinity (quantitative difference), or the various classes and subclasses of Ig might bind to different FcR (qualitative difference) (47-49). Among FcR-positive lymphocytes, the effector K cells were shown to be a minority subpopulation with a very high affinity for IgG (50). In studies on an FcR-positive murine macrophage line, only IgG2a antibodies were found to be cytophilic (51). Further, an important distinction was made recently between $T_{\cdot G}$ cells bearing an FcR for IgG, and $T_{\cdot M}$ cells bearing an FcR for IgM (52). IgM bound to $T_{\cdot M}$ cells can induce ADCC reactions (53, 54), a finding that might also play a role in NK. In our own experiments, a variety of types of complexes seem able to compete for the FcR responsible for ADCC against antibody-coated Chang cells (Table 1). FcR2 may bind only to certain types of complexes and this could account for the ability of some but not all complexes to inhibit NK.

A major problem with this model is that there are insufficient data in the literature concerning heterogeneity of FcR for binding of cytophilic antibodies. Tightly bound, non-IgG, cytophilic antibodies on NK cells is one interesting possiblity. Such non-IgG might obviate most of our criticisms against the arming of NK, and explain the results obtained with protein A. However, it is rather unlikely that cytophilic IgM antibodies are involved, since $T_{\cdot G}$ and $T_{\cdot M}$ are thought to be in different, nonoverlapping lymphocyte subsets (52), whereas NK cells certainly bear FcR for IgG (8). Until more information on these points becomes available, other models should be considered that do not rely on more than one FcR.

MODEL NO. 2: NK DUE TO A SPECIFIC MEMBRANE RECEPTOR ON FcR-POSITIVE CELLS

This model has been developed because our trypsinization and treatment with protein A suggest a difference between NK and ADCC, and because neither the EA- and Ig-inhibition experiments nor the available data in the literature provide conclusive evidence that NK is dependent on Ig. This model proposes that there are trypsin-sensitive membrane receptors on NK cells with some specificity for target cells. Only limited efforts have been made to characterize a cell surface receptor on NK cells or to demonstrate the physical adherence of effector cells to target cells. Monolayer adsorption experiments might be of some use. We have found that the G-11 (derived from breast carcinoma) or E-14 (derived from squamous cell carcinoma of lung) monolayer cell lines, which have been found to cross react with K-562 (55; Ortaldo et al., in preparation), could absorb a significant portion of NK activity. This is consistent with, but not proof of, a receptor for NK activity. In these experiments, ADCC activity against antibody-coated Chang liver cells was removed to the same extent as NK activity (Bonnard et al., in preparation). These preliminary results indicate that NK cells are also K cells and that at least an appreciable proportion of K cells have NK activity. However, in this model the NK receptors and the Fc receptors for ADCC activity are different and NK activity is not dependent on Ig. Binding of Ig to effector cells as in model no. 1 (specific antibodies to the target cells on FcR2), or no. 3 (triggering metabolic effects by the Ig-FcR interaction), or to cell-bound lymphotoxin as to be proposed elsewhere (14), would play only a supplementary or augmenting role. The data discussed regarding increased activity in the presence of various sera, or decreased activity upon exposure to antiglobulin reagents, would be related only to a secondary role of Ig. Model no. 2 and tells very little about the actual mechanisms of target cell killing. Model no. 3 and the model related to lymphotoxins (14) offer some alternative possibilities.

MODEL NO. 3: IgG BOUND TO FcR TURNS ON OR AUGMENTS MECHANISMS OF NK REACTIVITY

Some data suggest that Ig is involved in NK, but there is not yet enough evidence to decide whether there are natural antibodies with specificity for the target cells (e.g., model no. 1). Accordingly, it seems worthwhile to consider the possibility that binding of Ig leads to some activation of NK cells. The binding of IgG to the FcR on granulocytes has been found to account for the release of lysosomal enzymes (56, 57). Suppressor cells (52) and mitogenesis may also be activated by IgG-FcR interactions. This model postulates that native Ig or immune complexes without specificity for the target cells might nonetheless activate or augment NK. Linkage of NK cells and target cells could occur through specific membrane receptors (model no. 2). Recent reports on lymphotoxins have shown that these molecules could also establish a selective bond between effector cells and target cells (14). In contrast, in ADCC reactions, IgG antibody serves as the bridge between effector cells and target cells, and also the Fc portion of IgG bound to its target antigen may be a far better ligand and activator of FcR-positive K cells than would be the Fc or free, unbound IgG. These conditions, peculiar to ADCC, may account for this reaction being generally more effective than NK, when measured on the same target cells. Binding of NK cells to their target cells through a specific membrane receptor (model no. 2) might not by itself provide the same strong activation of metabolic cytolytic processes. This model also provides some explanation for our results in Table 1, if we consider that a variety of immune complexes, even unrelated to the target cells, could act as "activators" of NK, rather than as inhibitors.

CONCLUDING REMARKS

Several tentative conclusions can be drawn from our data and those in the literature on the role of Ig in NK. It has been difficult to demonstrate effects on NK by Ig; the precise nature of the reagents used and the exact conditions of testing bear considerably on the results. We have seen some inconsistent boosting effects on NK by allogeneic or syngeneic sera, but it is not possible at present to ascribe these effects to IgG antibodies specific for the target cells. We have been unsuccessful in showing that effector cells armed by cytophilic IgG, could account for all of the observed NK reactions. In fact, our experiments with immune complexes (Table 1) speak strongly against a central role of FcR-bound Ig in NK. Further, we have not been able to consistently "rearm" trypsinized NK cells with serum (Table 2), and thus have not been able to reproduce the results of Takasugi et al., (25-27). Rather, trypsin seemed to cleave off or modify some essential membrane structure needed for NK (8, 14). However, since it is clear that all NK cells bear an FcR and can function as K cells, it is likely that in the presence of appropriately bound specific antitarget cell antibodies, ADCC would occur in addition to NK. In model no. 2, we proposed this as an auxillary mechanism for cells that would have specific membrane receptors for target cells of NK. In model no. 3, we ascribed much more importance to Ig and immune complexes, in that they could trigger the metabolic events leading to lysis. It would seem reasonable to assume some overlapping between NK and ADCC reactions, and this may account in part for the divergent results of different groups at this conference owing to differences in reagents and cytotoxicity systems.

References

1. Bonnard, G. D., and West, W. H. *In* "Immunodiagnosis of Cancer" (R. B. Herberman and K. R. McIntire, eds.). Marcel Dekker, New York, in press.
2. Bonnard, G. D., West, W. H., Ortaldo, J. R., Kay, H. D., Cannon, G. B., and Herberman, R. B. *In* "Progress in Immunology III" (D. S. Nelson, ed.), p. 547. Elsevier-North-Holland, New York, 1978.
3. Herberman, R. B., and Holden, H. T. *Adv. Cancer Res.* 27, 305, 1978.

4. Herberman, R. B., Nunn, M. E., Holden, H. T., and Lavrin, D. H. *Int. J. Cancer* **16**, 230, 1975.
5. Kiessling, R., Petranyi, G., Karre, K., Jondal, M., Tracey, D., and Wigzell, H. *J. Exp. Med.* **143**, 772, 1976.
6. Kiessling, R. *et al. In* "Cytotoxic Cell Interactions and Immunostimulation (G. Riethmüller and G. Cudkowicz, eds.). Academic Press, New York, 1979.
7. West, W. H., Cannon, G. B., Kay, H. D., Bonnard, G. D., and Herberman, R. B. *J. Immunol.* **118**, 355, 1977.
8. Kay, H D., Bonnard, G. D., West, W. H., and Herberman, R. B. *J. Immunol.* **118**, 2058, 1977.
9. West, W. H., Boozer, R. B., and Herberman, R. B. *J. Immunol.* **120**, 90, 1978.
10. Herberman, R. B., Holden, H. T., West, W. H., Bonnard, G. D., Santoni, A., Nunn, M. E., Kay, H. D., and Ortaldo, J. R. *In* "Proceedings of International Symposium on Tumor-Associated Antigens and Their Specific Immune Response," Milan, Italy, June 6-8, 1977, in press.
11. Herberman, R. B., Bartram, S., Jr., Haskill, S., Nunn, M., Holden, H. T., and West, W. H. *J. Immunol.* **119**, 322, 1977.
12. Ortaldo, J. R., Bonnard, G. D., and Herberman, R. B. *J. Immunol.* **119**, 1351, 1977.
13. Herberman, R. B., Nunn, M. E., and Holden, H. T. *J. Immunol.* **121**, 304, 1978.
14. Bonnard, G. D. Submitted for publication.
15. Pape, G. R., Troye, M., and Perlmann, P. *J. Immunol.* **118**, 1919, 1977.
16. Pape, G. R., Troye, M., and Perlmann, P. *J. Immunol.* **118**, 1925, 1977.
17. Perlmann, P. *et al. In* "Cytotoxic Cell Interactions and Immunostimulation" (G. Riethmüller and G. Cudkowicz, eds). Academic Press, New York, 1979.
18. Troye, M., Perlmann, P., Pape, G. R., Spiegelberg, H. L., Näslund, I., and Giflöf, A. *J. Immunol.* **119**, 1061, 1977.
19. Troye, M. *et al. In* "Cytotoxic Cell Interactions and Immunostimulation" (G. Riethmüller and G. Cudkowicz, eds.). Academic Press, New York, 1979.
20. Rieber, E. P. *et al. In* "Cytotoxic Cell Interactions and Immunostimulation" (G. Riethmüller and G. Cudkowicz, eds.). Academic Press, New York, 1979.
21. Pape, G. R. *et al. In* "Cytotoxic Cell Interactions and Immunostimulation" (G. Riethmüller and G. Cudkowicz, eds.). Academic Press, New York, 1979.
22. Kumagai, K., Abo, T., Sekizawa, T., and Sasaki, M. *J. Immunol.* **115**, 982, 1975.
23. Lobo, P. I., and Horwitz, D. A. *J. Immunol.* **117**, 939, 1976.
24. Winchester, R. J., Fu, S. M., Hoffman, T., and Kunkel, H. G. *J. Immunol.* **114**, 1210, 1975.
25. Koide, Y., and Takasugi, M. *J. Natl. Cancer Inst.* **59**, 1099, 1977.
26. Takasugi, M. *et al.* This Volume.
27. Takasugi, J., and Takasugi, M. *Transplantation* **24**, 325, 1977.
28. Perlmann, P., Perlmann, H., and Biberfeld, P. *J. Immunol.* **108**, 558, 1972.
29. Imir, T., Saksela, E., and Makela, O. *J. Immunol.* **117**, 1938, 1976.
30. Santoli, D., Trinchieri, G., Zmijewski, C. M., and Koprowski, H. *J. Immunol.* **117**, 765, 1976.
31. Peter, H. H., Pavie-Fischer, J., Fridman, W. H., Aubert, C., Cesarini, J., Roubin, R., and Kourilsky, F. M. *J. Immunol.* **115**, 539, 1975.
32. Kiuchi, M., and Takasugi, M. *J. Natl. Cancer Inst.* **56**, 575, 1976.
33. Parillo, J. E., and Fauci, A. S. *Clin. Exp. Immunol.* **31**, 116, 1978.
34. Stejskal, V., Pape, G. R., and Perlmann, P. *In* "Regulatory Mechanisms in Lymphocyte Activation" (D. O. Lucas, ed.), p. 548. Academic Press, New York, 1977.
35. Peter, H. H., Eife, R. F., and Kalden, J. R. *J. Immunol.* **116**, 342, 1976.
36. Saal, J. G., Rieber, E. P., Riethmüller, G., and Hadam, M. *Z. Immun.-Forsch.* **152**, 110, 1976.
37. Ortaldo, J. R., Kay, H. D., and Bonnard, G. D. *In* "Regulatory Mechanisms in Lymphocyte Activation" (D. O. Lucas, ed.), p. 542. Academic Press, New York, 1977.

38. Trinchieri, G., Santoli, D., Zmijewski, C. M., and Koprowski, H. *Transplant. Proc.* 9, 881, 1977.
39. Trinchieri, G. *et al.* This volume.
40. Oehler, J. R., Lindsay, L. R., Nunn, M. E., Holden, H. T., and Herberman, R. B. *Int. J. Cancer* 21, 210, 1978.
41. Oehler, J. R., and Herberman, R. B. *Int. J. Cancer* 21, 221, 1978.
42. Berg, K., Ogburn, C. A., Paucker, K., Mogensen, K. E., and Cantell, K. *J. Immunol.* 114, 640, 1975.
43. Horwitz, D. A. *Fed. Proc.* 35, 473, 1976.
44. Nelson, D. L., Bundy, B. M., Pitchon, H. E., Blaese, R. M., and Strober, W. *J. Immunol.* 117, 1472, 1976.
45. Cordier, G., Samurut, C., and Revillard, J. P. *J. Immunol.* 119, 1943, 1977.
46. Ziegler, H. K., and Henney, C. S. *J. Immunol.* 115, 1500, 1975.
47. Froland, S. S., Michaelsen, T. E., Wisloff, F., and Natvig, J. B. *Scand. J. Immunol.* 3, 509, 1974.
48. Dickler, H. B. *Adv. Immunol.* 24, 167, 1976.
49. Winfield, J. B., Lobo, P. I., and Hamilton, M. E. *J. Immunol.* 119, 1778, 1977.
50. Cordier, G., Samurut, C., and Revillard, J. P. *In* "Leukocyte Membrane Determinants Regulating Immune Reactivity" (V. P. Eijsvoogel, D. Roos, and W. P. Zeijlemaker, eds.), p. 619. Academic Press, New York, 1976.
51. Walker, W. S. *J. Immunol.* 119, 367, 1977.
52. Moretta, L., Webb, S. R., Grossi, C. E., Lydyard, P. M., and Cooper, M. D. *J. Exp. Med.* 146, 184, 1977.
53. Wahlin, B., Perlmann, H., and Perlmann, P. *J. Exp. Med.* 144, 1375, 1976.
54. Fuson, E. W., and Lamon, E. W. *J. Immunol.* 118, 1907, 1977.
55. Ortaldo, J. R., Oldham, R. K., Cannon, G. B., and Herberman, R. B. *J. Natl. Cancer Inst.* 59, 77, 1977.
56. Goldstein, I. M., Kaplan, H. B., Radin, A., and Frosch, M. *J. Immunol.* 117, 1282, 1976.
57. Henson, P. M., and Oades, Z. G. *J. Clin. Invest.* 56, 1053, 1975.

OPPOSING EFFECTS OF INTERFERON ON NATURAL KILLER AND TARGET CELLS

Giorgio Trinchieri and Daniela Santoli

Unsensitized human lymphocytes from normal individuals are spontaneously cytotoxic *in vitro* for target cell lines (1-5). The effector cells (natural killer cells) have been identified as Fc receptor-bearing lymphocytes that have no surface immunoglobulin and that do not form high-affinity rosettes with sheep red blood cells (2, 3). Similar spontaneous cytotoxic activity has been described for mouse (6-8) and rat lymphocytes (9). Evidence that the major histocompatibility complex plays a role in the genetic control of natural killer cell activity has been obtained both in man (4, 5, 10) and in mice (11). We have recently described that certain cell lines, when cultured *in vitro* with human or mouse lymphocytes, induce the production of interferon (12, 13). While normal fibroblasts are unable to induce interferon, most of the tumor-derived cell lines and of human lymphoid cell lines transformed by Epstein-Barr virus induce human lymphocytes to produce high levels of interferon starting a few hours after mixture. Interferon has been reported to increase both specific and nonspecific cell-mediated cytotoxicity of mouse lymphocytes (14, 15). In this communication two antagonistic effects of interferon on human natural killer cell-mediated cytotoxicity are considered: (a) interferon strongly enhances the activity of human cytotoxic lymphocytes; and (b) it abolishes almost completely the susceptibility of certain target cell lines to lysis (16).

Interferon was induced in human lymphocytes either by infection with Newcastle disease virus (NDV) or by cultivation with a human rhabdosarcoma-derived (RDMC) cell line (12). Antiviral titers of interferon are expressed as the reciprocal of the dilution inhibiting 50% of the cytopathic effect of vesicular

75

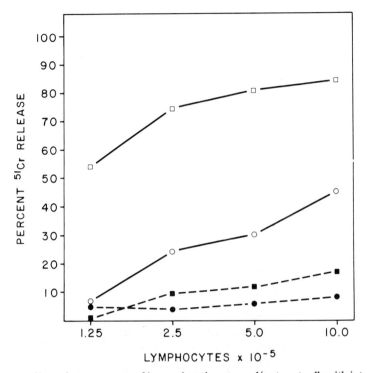

FIG. 1.   Effect of pretreatment of human lymphocytes and/or target cells with interferon. Monolayers of ^{51}Cr-labeled fetal skin fibroblasts were prepared in the wells of flat bottom microplates (Falcon 3040) (approximately 4×10^4/well) and incubated for 18 hr at 37°C with or without 10^3 units of interferon. Human peripheral blood lymphocytes were isolated on a Ficoll-Hypaque cushion and incubated for 18 hr at 37°C, 5×10^6/ml in RPMI 1640 medium supplemented with 10% fetal bovine serum and containing or not containing 10^3 units of interferon. Both lymphocytes and target fibroblasts were washed three times before testing. Four different dilutions of lymphocytes were added in triplicate to the well containing the target cells. Specific ^{51}Cr release and 50% lytic units against the treated and untreated target cells were determined as previously described (4, 15). The standard deviation of the experimental points was always less than 2%. □——□, untreated target cells, interferon-treated lymphocytes; ○——○, untreated target cells, untreated lymphocytes; ■——■, interferon-treated target cells, interferon-treated lymphocytes; ●——●, interferon-treated target cells, untreated lymphocytes.

stomatitis virus on monolayers of human fetal skin fibroblasts. One antiviral unit is equivalent to approximately 1 reference unit of the NIH Human Reference Interferon G-023-901-527.

Human peripheral blood lymphocytes were incubated for 18 hr at 37°C with diluted interferon preparations. The spontaneous cytotoxicity of the lymphocyte preparations was tested on monolayers of ^{51}Cr-labeled target cells as previously described (4). When normal human fibroblasts were used as target cells, the cytotoxic efficiency of lymphocytes incubated with interferon was 5- to 10-fold greater than that of lymphocytes tested immediately after separation or of those cultured for 18 hr at 37°C without interferon (Figs. 1 and 2a). About 10 antiviral units of interferon were required to effect a 2-fold increase in the cytotoxic efficiency of human lymphocytes (Fig. 2a). Cell-separation experiments have shown that the cytotoxic effector cells enhanced by interferon are contained in the same fractions as the natural killer cells, which suggested to us that interferon enhances natural killer cell activity rather than generates a new effector cell population.

Monolayers of human fibroblasts were incubated with different dilutions of interferon preparations and then tested as target cells: the treatment with interferon rendered them almost completely resistant to the cytotoxic effect of both nonenhanced and enhanced natural killer cells (Figs. 1 and 2b). Treatment with approximately 1 antiviral unit of interferon was required to inhibit lysis of human fetal fibroblasts by 50% (Fig. 2). The induction of resistance to lysis requires a preincubation of several hours and is prevented by concomitant treatment of target cells with cycloheximide (50 μg/ml) or actinomycin D (1 μg/ml). That interferon does not induce resistance in the presence of protein or RNA synthesis inhibitors suggests that, in analogy with the induction of antiviral status, this effect of interferon is mediated through derepression of the synthesis of an effector cell protein. The infection of fibroblasts with viruses such as influenza and vaccinia rendered them relatively insensitive to this interferon effect, possibly because the virus infection inhibited protein synthesis in the cells. The interferon-treated target cells are not as efficient in a competitive inhibition test as untreated cells. This suggests that these target cells have lost some "antigen" or "receptor" necessary for interaction with killer cells.

The mechanism inhibiting lysis of target cells seems specific for natural killer cells. The cytotoxicity mediated by PHA-stimulated lymphocytes is only partially affected, and mouse cytotoxic T cells sensitized against human cells kill interferon-treated and control human cells with the same efficiency.

Target cell lines other than fibroblasts have also been tested for induction of resistance to cytotolysis upon treatment with interferon. The results obtained with transformed or tumor-derived cell lines varied; some cell lines were quite susceptible, but others, especially tumor-derived lines, were almost completely resistant to the protective effect of interferon (Table 1). The susceptibility of the cell lines to the antiviral effect of interferon significantly correlated with the induction of resistance to cytotoxic cells (Table 1).

In most of the experiments presented, crude preparation of interferon were used, and those molecules that mediate the two effects on cytotoxicity have not been identified. All the evidence however, suggests that these two activities may be ascribed to interferon molecules as such. The effects on cell-mediated cytotoxicity have been observed with interferon preparations obtained by two different methods, whereas mock interferon preparations did not produce the effects. The titer of the activity on lymphocytes and on target cells in many interferon preparations tested was significantly correlated with the antiviral titer (Fig. 2). The components in the interferon preparations responsible for the activity on cyto-

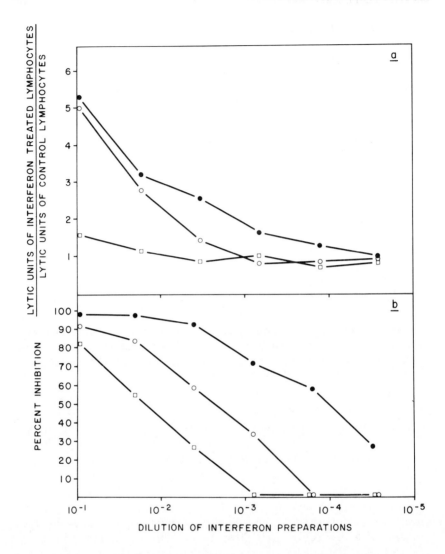

FIG. 2. Effect of three different interferon preparations on enhancement of cytotoxicity (a) and induction of resistance in target cells (b). Interferon was induced by culturing lymphocyte preparations from different donors on monolayers of RDMC cells[14]. Three preparations were used containing the following antiviral units: $\circ$—$\circ$, 10^4; $\circ$—$\circ$, 10^3; and $\square$—$\square$, 0.64×10^2. (a) 2×10^5 lymphocytes were incubated at $37°$C in triplicate wells of round bottom microplates (Limbro) in 200 μl of medium containing different dilutions of the interferon preparations. After 18 hr of incubation, 100 μl of supernatent was removed from each well and replaced with 100 μl of medium containing 4×10^4 ^{51}Cr-labeled RDMC target cells. After 8 hr, 100 μl of supernatents were collected from each well and the specific ^{51}Cr release was determined as previously described. 50% lytic units were computed on the basis of a standard curve obtained by testing several dilutions of lymphocytes stimulated with the highest interferon concentration. (b) Monolayers of ^{51}Cr-labeled fetal skin fibroblasts in flat bottom microplate were incubated 18 hr at $37°$C with the different dilutions of

TABLE 1. Susceptibility to Induction of Resistance
to Natural Killer Cell-Mediated Cytotoxicity and to Inhibition
of Viral Replication in Human Cell Lines Treated with Interferon

Cell line	Origin	% Inhibition cytotoxicity[a]	Inhibition of viral replication[b]
FS-1	Fetal skin fibroblasts	92	13
WI38	Fetal lung fibroblasts	62	10
MRC5	Fetal lung fibroblasts	40	6
Pa	Skin fibroblasts	86	12
LR-1	Newborn brain	82	11
LN-SV	Skin, SV40-transformed	77	10
S1054TR	Brain, SV40-transformed	20	1
W18, VA2	Lung, SV40-transformed	75	7
SW690	Melanoma	75	8
SW691	Melanoma	13	1
SW480	Colorectal carcinoma	24	1
D98 (HeLa)	Cervical carcinoma	30	1
HT1080	Fibrosarcoma	15	2
RDMC	Rhabdomyosarcoma	18	1

[a] Figures refer to the average % inhibition of cytotoxicity in target cells treated with interferon and tested against preparations of lymphocytes from two different donors. Monolayers of the different cell lines in the wells of flat bottom microtiter plates were incubated 18 hr with or without 10^3 antiviral units of interferon and labeled with ^{51}Cr. Dilutions of human lymphocytes (preincubated 18 hr with 10^3 units of interferon) were added to triplicate wells, and 50% lytic units on treated and untreated target cells were determined as previously described [refs. (4, 17)].

[b] Log_2 of the reciprocal of the dilution of a leukocyte interferon preparation (10^4 antiviral reference units) inhibiting 50% of the cytopathic effect of vesicular stomatitis virus on monolayers of the different cell lines.

toxicity were, like interferon, species-specific, destroyed by trypsin treatment, and resistant to pH 2. The single component active on cytotoxicity and responsible for the antiviral activity was eluted from a Sephadex G-100 column as a major peak corresponding to a molecular weight of approximately 25,000. Moreover, the protective effect of interferon preparations on target cells is active only on cells sensitive to the antiviral effect of interferon and it is similarly mediated through an intracellular mechanism that requires RNA and protein synthesis.

When fresh human lymphocytes are mixed *in vitro* with certain tumor-derived or virus-infected target cells, they produce interferon in the culture (12). After the first 4-5 hr of incubation, this endogenous interferon enhances the activity

Fig. 2 (continued)

interferon and then tested as target cells with human lymphocytes (cultured for 18 hr at $37°$C in the presence of 10^3 antiviral units of interferon). Specific ^{51}Cr release and 50% lytic units were determined in an-18 hr test, as previously described (4, 15). The % inhibition of cytotoxicity was computed according to the formula:

$$\% \text{ inhibition} = \frac{\text{lytic units on interferon-treated target cells}}{\text{lytic units on control target cells}} \times 100$$

of the natural killer cells present in the culture. It can be demonstrated that the interferon enhancement of spontaneous cell-mediated cytotoxicity is responsible for more than 80% of the cytotoxicity observed in an 18 hr test against target cells able to induce interferon (16).

Since titers of interferon equal to, or higher than those required to stimulate killer cells and to protect normal fibroblasts are present *in vivo* during viral infections or other inducing conditions, it is conceivable that *in vivo* interferon may activate a selective defense mechanism of cell-mediated cytotoxicity (Fig. 3). A significant increase in natural killer cell activity has been observed in mice injected with different types of viruses, BCG, and tumor cell lines (18), and it is possible that this effect is mediated through the production of interferon. If the protection by interferon of normal, but not virus-infected or tumor cells, operates *in vivo*, it might explain the paradoxical existence *in vivo* of natural killer cells that *in vitro* can nonspecifically lyse almost any target cells, including also normal autologous cells (4). Acting on lymphocytes, interferon can stimulate

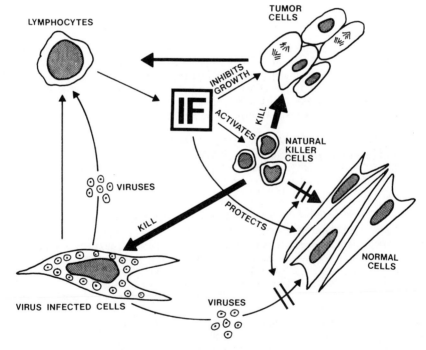

FIG. 3. Summary of the observed effects of Interferon (IF) on the activity of natural killer cell *in vitro*. Viruses, virus-infected cells, most tumor-derived cell lines, and other interferon inducers stimulate lymphocytes to produce interferon. Interferon inhibits growth of rapidly dividing cells, inhibits viral replication in susceptible cells and enhances the activity of natural killer (NK) lymphocytes. NK cells lyse nonspecifically *in vitro* any target cell lines. Interferon, however, is able to render normal cells (fibroblasts) and some tumor cell lines resistant to the cytotoxicity of NK cells, whereas virus-infected cells and most tumor-derived cell lines are not protected.

very effective nonspecific cytotoxic cells, but it can also selectively protect normal cells, directing the defense mechanism selectively against tumor cells or virus-infected cells. However, in some instances this protective effect could also represent an escape mechanism for tumor cells (e.g., the melanoma cell line SW690, Table 1) that have maintained susceptibility to interferon.

Acknowledgments

This work was supported in part by USPHS research grant CA-20833, CA-10815, CA-43882 from the NCI, NS-11036 from the NINCDS, by grant IM-88 from the American Cancer Society, and by the National Multiple Sclerosis Society. We acknowledge Drs. Albert Leibovitz, Barbara Knowles, and Zenon Steplewski for cell lines and Ms. Nancy Seitz for technical assistance.

References

1. Takasugi, M., Mickey, M. R., and Terasaki, P. I. *Cancer Res.* **33**, 2898-2902, 1973.
2. Jondal, M., and Pross, H. *Int. J. Cancer* **15**, 596-605, 1975.
3. Peter, H., Pavie-Fischer, J., Fridman, W., Aubert, C., Cesarini, P., Roubin, R., and Kourilsky, F. *J. Immunol.* **115**, 539-548, 1975.
4. Santoli, D., Trinchieri, G., Zmijewski, C. M., and Koprowski, H. *J. Immunol.* **117**, 765-770, 1976.
5. Trinchieri, G., Santoli, D., Zmijewski, C. M., and Koprowski, H. *Transplant Proc.* **9**, 881-884, 1977.
6. Herberman, R. B., Nunn, M. E., and Lavrin, D. H. *Int. J. Cancer* **16**, 216-226, 230-239, 1975.
7. Kiessling, R., Klein, E., and Wigzell, H. *Eur. J. Immunol.* **5**, 112-117, 1975.
8. Kiessling, R., Klein, E., Pross, H., and Wigzell, H. *Eur. J. Immunol.* **5**, 117-121, 1975.
9. Nunn, M. E., Djeu, J. Y., Glaser, M., Lavrin, D. H., and Herberman, R. B. *J. Natl. Cancer Inst.* **56**, 393-399, 1976.
10. Petranyi, G. G., Benczur, M., Onody, C. E., and Hollan, S. R. *Lancet* **1**, 736, 1974.
11. Petranyi, G. G., Kiessling, R., Povey, S., Klein, E., Herzenberg, L., and Wigzell, H. *Immunobiogenetics* **3**, 15-28, 1976.
12. Trinchieri, G., Santoli, D., and Knowles, B. *Nature* **270**, 611-613, 1977.
13. Trinchieri, G., Santoli, D., Dee, R., and Knowles, B. B. *J. Exp. Med.* **147**, 1299-1313, 1978.
14. Chernyakhovskaya, I. Y., Slavina, E. G., and Svet-Moldavsky, A. J. *Nature (London)* **228**, 71-72, 1970.
15. Lindahl, P., Leary, P., and Gresser, I. *Proc. Natl. Acad. Sci.* **69**, 721-725, 1972.

16. Trinchieri, G., and Santoli, D. *J. Exp. Med.* **147**, 1314-1323, 1978.
17. Trinchieri, G., Baumann, P., Demarchi, M., and Tokes, Z. *J. Immunol.* **115**, 249-255, 1975.
18. Herberman, R. B., Nunn, M. E., Holdenti, T., Staal, S., and Djeu, J. Y. *Int. J. Cancer* **19**, 555-564, 1977.

SPECIFICITY OF NATURAL CYTOTOXIC CELLS DETERMINED BY ANTIBODY

Mitsuo Takasugi and Donna Akira

Natural or spontaneous cell-mediated cytotoxicity (NCMC) was originally observed as an "interfering" background reaction by cells of normal individuals (1, 2). Isolated lymphocyte suspensions of such individuals were tested against cultured tumor target cells in the search for tumor specificity. It is now well established that virtually all individuals exhibit this type of reactivity. When natural cytotoxicity is assayed on a variety of target cells, a significant component of the reaction appears to be nonselective (3). Yet, the overall reactions against each target were highly specific (4).

The specificity of NCMC against each target was investigated in the cross-competition assay by specific inhibition of cytotoxicity against several ^{51}Cr-labeled targets by the same set of unlabeled targets. NCMC against each target was selectively inhibited by the same competing target, indicating that some cells in the effector suspension detected antigens distinct for each of the targets. Thus, nonselective cytotoxicity can be derived from specific reactions, i.e., specific activity against a different antigen on each of the targets or against an antigen shared by all.

Although some differences in opinion still exist concerning the characteristics of the effector cell for NCMC, there is emerging agreement that in human systems it carries Fc receptors for IgG molecules (5-9). These receptors appear to have an important role in NCMC reactions since aggregated human IgG and antigen-antibody complexes inhibit natural cytotoxicity. A strong similarity also exists between the effector cells for NCMC and antibody-dependent cell-mediated cytotoxicity (ADCC) with the exception that effector cells for NCMC, but not ADCC, are sensitive to proteases (10, 11).

The effector cells for NCMC have been reported to carry immunoglobulins attached to their surface through Fc receptors (12, 13). These antibodies may be removed by proteases, or low pH, and reconstituted by incubating the effector cell suspension in serum (14). NCMC activity accompanies the presence of antibodies on effector cell surfaces during this removal and reconstitution. If IgG antibodies attach to effector cells through Fc during short periods of incubation in serum, lymphocytes isolated from the peripheral blood should already possess attached natural antibodies from the circulation.

MATERIALS AND METHODS

The cytotoxic test employing ^{51}Cr-release has been described (4, 11). Briefly, lymphocytes isolated from peripheral blood by Ficoll-Hypaque discontinuous gradient centrifugation were tested with cultured target cells for 16 hr at an effector-target ratio of 40:1 in the wells of a "V" microtiter plate (Cooke Engineering, Alexandria, Virginia). The test was terminated by spinning the plates and removing 0.1 ml supernatant from each well for counting. All tests and controls were performed in triplicate.

$$\% \text{ cytotoxicity} = \frac{\text{test} - \text{minimum release}}{\text{maximum} - \text{minimum release}} \text{ X } 100$$

nonselective cytotoxicity = effector cell av. + target cell av. − overall av.

selective cytotoxicity = observed − nonselective cytotoxicity

Effector lymphocytes were treated with 0.25% trypsin at a concentration of 3-4 x 10^6/ml for 30 min at 37°C. The cells were washed thoroughly, and either tested directly or incubated in 0.2 ml serum for 3 hr at 37°C and tested.

Serum from a normal individual was divided into 0.2 ml aliquots and absorbed three times with 3 x 10^6 cultured target cells for 30 min each. Antibodies were also eluted from the first absorption by washing the absorbing cells and treatment with 0.1 ml, pH = 4 acetate buffer. The eluate was neutralized to pH = 7 with approximately 0.2 ml NaOH in media.

RESULTS

Freshly isolated lymphocytes were treated with trypsin to remove antibodies and to reduce preexisting natural cytotoxicity. They were then reconstituted by incubation in normal serum and aliquots of this serum that had been absorbed with three separate cultured targets. Effector cells reconstituted with these sera were tested for selective cytotoxic activity against the same three targets.

Table 1 gives the results in counts per minute (part I), % cytotoxicity before (part 2), and after differentiation of selective from nonselective cytotoxicity by application of the interaction analysis (part 3). Cytotoxicity was selectively lower or negative for a given target when the effectors were reconstituted with serum absorbed by that target. Natural antibodies specific for the target were removed by the absorption.

Effectors were also reconstituted with antibodies eluted from the same absorbing target. The cell pellet following absorption of the serum was washed twice in Hank's balanced salt solution and the antibody eluted in acetate buffer at pH = 4. The eluate was neutralized and used to reconstitute effector cells. The results of

TABLE 1. Reconstitution of NCMC Effector Cells by Incubation in Serum Absorbed with Target Cells

Effector 158552 Trypsin	Incubated in serum absorbed with	Target cells (counts/min)		
		497	917	372
(1) +	–	566	352	2949
+	497	541	353	2974
+	917	560	319	2871
+	372	561	334	2808
Minimum release		490	315	2638
Maximum release		949	656	4530
Range		459	341	1892
(2)		% cytotoxicity		
+	–	16.6	10.9	16.4
+	497	11.1	11.1	17.8
+	917	15.3	1.1	12.3
+	372	15.5	5.6	8.9

(3)

Effector reconstituted with serum absorbed with		Cytotoxicity			Effector cell av.
497	Observed	11.1	11.1	17.8	12.3
	Nonselective	15.4	7.3	14.4	
	Selective	−4.3	3.8	3.4	
917	Observed	15.3	1.1	12.3	10.5
	Nonselective	13.6	6.3	12.6	
	Selective	1.7	−5.2	−0.3	
372	Observed	15.5	5.6	8.9	10.0
	Nonselective	13.1	5.0	12.1	
	Selective	2.4	0.6	−3.2	
	Target cell av.	14.0	5.9	13.0	10.9 Overall av.

tests with eluate-reconstituted cells are given in Table 2. Only the release in counts
per minute (part 1) and the selective results from the interaction analysis (part 2)
are shown. Effector cells reconstituted with antibody eluted from target cells
were selectively more active against that target indicating the existence of specific
antibody within the serum for each of the targets. There is a 50% probability of
achieving such a positive or negative selective result through absorption or elution.
Thus the chance of achieving negative selective results along the diagonal in absorption tests is $(\frac{1}{2})^3$, and for positive selective scores along the diagonal by elution
it is $(\frac{1}{2})^3$ for a combined probability of 0.02.

CONCLUDING REMARKS

The effector suspension armed with natural antibodies possesses the capacity
to react specifically with most cultured targets giving an appearance of nonselec-

TABLE 2. Reconstitution of NCMC Effector Cells by Incubation in Antibodies Eluted
from Target Cells Used for Absorption

(1)
Effector 158594

Antibodies eluted		Target cells (counts/min)		
Trypsin	from	497	917	372
+	497	305	450	729
+	917	292	432	722
+	372	276	442	714
+		253	385	743
Minimum release		250	384	612
Maximum release		516	665	1083
Range		266	281	471

(2)

Differentiation of selective cytotoxicity

Effector reconstituted with antibodies eluted from		% Cytotoxicity			Effector cell av.
497	Observed	15.8	17.1	23.4	18.8
	Nonselective	12.4	16.2	27.8	
	Selective	3.4	0.9	−4.4	
917	Observed	9.8	20.6	21.7	17.4
	Nonselective	11.0	14.8	26.4	
	Selective	−1.2	5.8	−4.7	
372	Observed	1.1	0.3	27.8	9.7
	Nonselective	3.3	7.1	18.7	
	Selective	−2.2	−6.8	9.1	
	Target cell av.	8.9	12.7	24.3	15.3 Overall av.

tiveness to these reactions. The effector cell itself appears to be nonselective; it is the natural antibodies to a wide spectrum of antigens that provide specificity to the reaction. It has been suggested that NCMC is directed against viral antigens and, indeed, natural antibodies to various viruses have been detected (15, 16). Antibody arming of effector cells could explain specific resistance to viruses where antibody alone was ineffective. Whether the specificities of natural antibodies acting in conjunction with effector cells against cultured human targets are really directed against viral antigens on such cells remains to be determined.

Each individual possesses a different spectrum of natural antibodies, resulting in different patterns of cytotoxicity against targets when their effector cells are reconstituted with corresponding antibodies. A more thorough investigation of natural cytotoxicity in man manifesting specific cytotoxic activity against a multiplicity of determinants will lead to a fuller understanding of host defenses and surveillance.

The apparent nonselective reactivity of NCMC for cultured target cells is the sum of multiple specific reactions. A subclass of lymphocytes with Fc receptors interacts with circulating natural antibodies that confer specificity. The role of antibodies in NCMC was demonstrated by the loss and recovery of cytotoxic activity with removal and replacement of antibodies during trypsinization of effector cells followed by their incubation in serum. When effector cell reconstitution was sought with absorbed serum, or with antibodies eluted from the absorbing target cells, reactivity was selectively lost or regained, thus affirming a decisive role for natural antibodies.

Acknowledgment

This work was supported in part by Public Health Service contracts NO1-CP-43211 from the Virus Cancer Program of the Division of Cancer Cause and Prevention, and CB-74133 from the Tumor Immunology Program Section, National Cancer Institute.

References

1. Takasugi, M., Mickey, M. R., and Terasaki, P. I. *Cancer Res.* **33**, 2898, 1973.
2. Oldham, R. K., Siworski, D., McCoy, J. L., Plate, E. J., and Herberman, R. B. *J. Natl. Cancer Inst. Mono.* **37**, 49, 1973.
3. Takasugi, M., and Mickey, M. R. *J. Natl. Cancer Inst.* **57**, 255, 1976.
4. Takasugi, M., Koide, Y., Akira, D., and Ramseyer, A. *Int. J. Cancer* **19**, 291, 1977.
5. Kiuchi, M., and Takasugi, M. *J. Natl. Cancer Inst.* **56**, 575, 1976.

6. Hersey, P., Edwards, A., Edwards, J., Adams, E., Milton, G. W., and Nelson, P. S. *Int. J. Cancer* **16**, 173, 1975.
7. Peter, H. H., Pavie-Fischer, J., Fridman, W. H., Aubert, C., Cesarini, J. P., Roubin, R., and Kourilsky, F. M. *J. Immunol.* **115**, 539, 1975.
8. West, W. H., Cannon, G. B., Kay, H. D., Bonnard, G. D., and Herberman, R. B. *J. Immunol.* **118**, 355, 1977.
9. Bakacs, T., Gergely, P., Cornain, S., and Klein, E. *Int. J. Cancer* **19**, 441, 1977.
10. Koide, Y., and Takasugi, M. *J. Natl. Cancer Inst.* **59**, 1099, 1977.
11. Akira, D., and Takasugi, M. *Int. J. Cancer* **19**, 747, 1977.
12. Winchester, R. J., Fu, S. M., Hoffman, T., and Kunkel, H. G. *J. Immunol.* **114**, 1210, 1975.
13. Lobo, P. I., Westervelt, F. B., and Horwitz, D. *J. Immunol.* **114**, 116, 1975.
14. Kumagai, K., Abo, T., Sekizawa, T., and Sasaki, M. *J. Immunol.* **115**, 982, 1975.
15. Nowinski, R. C., and Koehler, S. I. *Science* **185**, 869, 1974.
16. Aoki, T., Liu, M., Walling, M. J., Bushar, G. S., Brandchaft, P. B., and Kawakami, T. G. *Science* **191**, 1180, 1976.

MANIPULATIONS OF NATURAL CYTOTOXICITY IN TUMOR PATIENTS VIA BCG

Johannes G. Saal, Gert Riethmüller, Martin R. Hadam,
Ernst P. Rieber, and Joachim M. Fleiner

Bacillus Calmette-Guerin (BCG) is widely used as an immunostimulant for therapy in various malignant diseases of man. However, a rational basis for an individual dosage and timing of BCG administration is still lacking. This may be one of the reasons for which its therapeutic efficacy is still being disputed.

Recently it has been described that natural cytotoxicity of lymphocytes is increased in mice chronically infected with BCG (1). Similarly, we have reported that during the course of monthly BCG administration spontaneous cytolytic activity of human circulating lymphocytes is characteristically affected, and often increased (2, 3). Since natural, alias spontaneous cytotoxicity, is regarded as a possible *in vitro* correlate of a nonspecific immunological defense mechanism against tumors *in vivo*, it was of interest to study the dose-effect relationship and kinetics of BCG application on this cytolytic reaction in tumor patients. Such an approach may help to guide individual immunotherapy and help to detect early the possible adverse effects of the immunostimulant on tumor growth that have been described (3-9).

89

MATERIALS AND METHODS

Patients

The study comprises 18 patients with malignant melanoma and 35 healthy blood donors. The diagnosis was confirmed by histology in all tumor patients. All patients were clinically free of malignant disease during this study after surgery of the primary tumor or tumor recurrences.

BCG-Therapy

Patients were treated with BCG-scarifications after extirpation of the primary tumor or after ablative surgery of tumor recurrences. For evaluation of dose-dependent effects of BCG, various concentrations of fresh BCG-F (Pasteur Institute, Paris, 10^8, 10^7, and 10^6 organisms) were applied in 0.1 ml/32 cm scarification as previously described (3).

Target Cells

Cell lines derived from human malignant melanoma and established in our laboratory were used as target cells: Mel-Ho-88, Mel-Ju-87, Mel-JuS-90, and Mel-Ei-78. The cells were propagated in RPMI 1640 (GIBCO) supplemented exclusively with 10% human AB-serum, antibiotics, and L-glutamine. The long-term breast carcinoma line SKBr3 as well as the bladder carcinoma line T24 were generously provided by Dr. J. Fogh (Sloan Kettering Institute, New York) and were cultured in MEM (GIBCO) supplemented with 15% fetal calf serum (GIBCO), L-glutamine, nonessential amino acids, and antibiotics. The same culture medium was used for the Hep-2 cell line derived from a human larynx carcinoma and obtained from Flow Laboratories, Bonn, Germany.

Effector Cells

Nonadherent lymphocytes (NAL) were isolated from pheripheral blood by Ficoll®Urovison® density gradient centrifugation. Adherent cells were removed by incubation on plastic petri dishes and by passage of the supernatant cells through a nylon wool column as previously described (10).

^{3} H-Proline Release Test

Spontaneous cytolytic activity of lymphocytes was measured as previously described (10). In brief, 5×10^6-1×10^7 tumor cells labeled for 24 hr with 50 μCi ^{3} H-proline (500-1000 mCi/mM) were distributed into Falcon Microtest-II Plates (1×10^4/well). Effector lymphocytes at various concentrations were added in RPMI 1640 medium supplemented with 10% pooled human AB-serum. After 40 hr of incubation at 37°C, the total contents of each well were harvested on serum pretreated glass filters by use of a harvesting machine. The radioactivity retained on the filters was determined in a LS-Counter. Cytolytic activity (CTL) was calculated according to the formula

$$\% \ CTL = 100 - (\frac{cpm \ test \ sample}{cpm \ medium \ control} x \ 100)$$

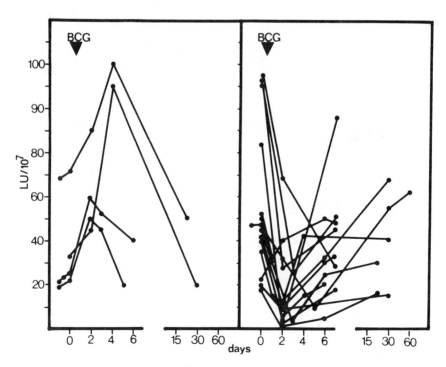

FIG. 1. Short-lasting effects of high-dose BCG (10^8 organisms/32 cm scarification) on CMC in 18 melanoma patients. CMC is expressed as lytic units/10^7 nonadherent lymphocytes (LU/10^7). Target cell: Mel-Ei-78.

Cytolytic activity was finally expressed in lytic units; one lytic unit (LU) being defined as the number of lymphocytes required to achieve 25% target cell lysis.

RESULTS

The effect of local application of BCG on the spontaneous cytolytic activity of lymphocytes (CMC) was investigated in 18 melanoma patients. In all cases the highest dose of BCG tested, namely, 10^8 organisms/32 cm scarification, induced systemic cytotoxicity responses that followed the characteristic time kinetics shown in Fig. 1. Eighty percent (14/18) of the patients reacted with an immediate short-lasting decrease of cytotoxicity, whereas only 20% (4/18) demonstrated a short-lasting increase in cytotoxicity. Both types of responses reached their maxima between day 2 and day 4 after BCG-scarification and leveled out within 8 to 10 days. Thereafter the cytolytic activity remained relatively constant until the next BCG treatment was applied. Since test plates with the same batch of ^{3}H-proline-labeled target cells were used for all assays performed during one week, alteration of target cells during prolonged culture could have caused the described cytolytic response patterns. This possibility was ruled out, however, by testing in parallel healthy individuals without BCG treatment on the same target cells. As shown in Fig. 2, their cytotoxic activity showed considerable short-term stability.

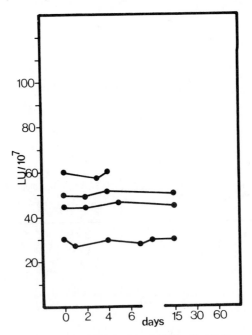

FIG. 2.   Short-term stability of CMC in four healthy blood donors without BCG tested under the same experimental conditions as the melanoma patients. CMC is expressed as lytic units/10^7 nonadherent lymphocytes (LU/10^7). Target cell: Mel-Ei-78.

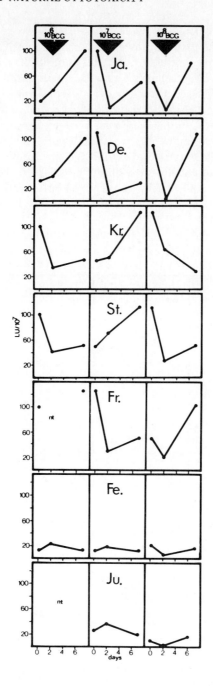

FIG. 3. Dose-dependent short-term BCG effects on CMC in seven melanoma patients tested. CMC is expressed as lytic units/10^7 nonadherent lymphocytes (LU/10^7). Target cell: Mel-Ei-78.10^6, 10^7, and 10^8 BCG organisms were applied at weekly intervals; nt = not tested.

In another experimental approach, the question was asked whether the different patterns of cytotoxic responses after administration of 10^8 BCG organisms were influenced by the dose of BCG. For this purpose, three doses, 10^6, 10^7, and 10^8 BCG organisms, were administered at weekly intervals to the same patient. This schedule was followed in single patients starting with either the high dose or the low dose. From the small number of patients studied so far, a distinct dose-dependent pattern of cytotoxic responses emerged. As can be seen from Fig. 3, application of 10^8 organisms/32 cm scarification resulted in a decrease of cytolytic activity in all patients tested. Lower BCG doses, however, were able to induce augmentation of cytotoxicity. Interestingly enough, there seems to be an optimal boosting dose that is individually different and varies from 10^6 organisms in the patients Ja, De, Fe, and probably patient Fr, to 10^7 organisms in the patients Kr, St, and Ju.

Considering the fact that certain melanoma patients analyzed reacted in a disease-related manner preferentially against melanoma target cells (3), it was of

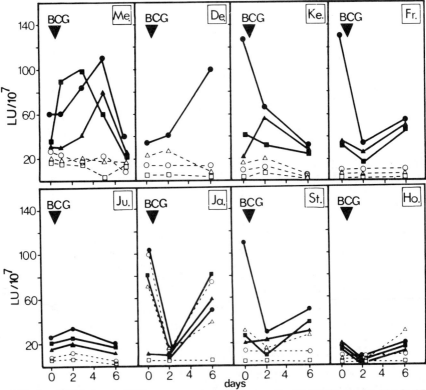

FIG. 4. BCG-induced CMC responses on melanoma cells (Mel-Ei-78, ●——●; Mel-Ju-87, ▲——▲, Mel-Ho-88, ■——■) and nonmelanoma cells (T24, ○——○; SKBr3, △——△; Hep-2, □——□). Eight patients were analyzed: 10^6 (De, St), 10^7 (Ja, Fr, Ju) or 10^8 (Ho, Ke, Me) BCG organisms were applied per scarification. CMC is expressed as lytic units/10^7 nonadherent lymphocytes (LU/10^7).

interest to know how BCG influenced the cytotoxic response against melanoma cells compared to other tumor target cells. For that reason, eight melanoma patients receiving BCG were tested on three melanoma and three nonmelanoma target cell lines. As expected from previous experience, melanoma patients tended to show a higher cytotoxicity against melanoma than against the nonmelanoma targets when tested prior to BCG treatment (Fig. 4). In addition, it was found that the individual response pattern was usually the same on all three melanoma target cells tested, except for patients Ke and St. However, the most interesting observation was the increased cytotoxic response on melanoma target cells not paralleled by a similar increase of cytotoxicity against nonmelanoma targets (see patients Me, De, and Ju). The fact that detection of decreased cytotoxicity was restricted to melanoma target cells may be explained by the very low pretreatment cytotoxicity against nonmelanoma cells. This explanation is supported by the observation of decreased cytotoxicity against other tumor cells in patient Ja; this is the only case with high pretreatment level of activity against nonmelanoma cells and decline of cytolytic activity.

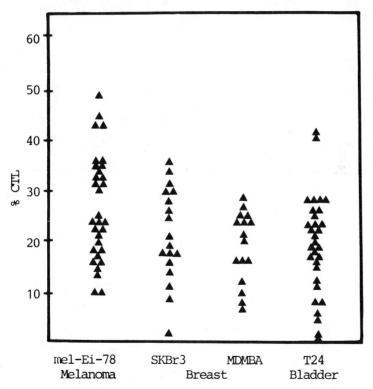

FIG. 5. Distribution of cytolytic activities of lymphocytes from 31 healthy donors on different tumor target cells. The cytolytic activity of nonadherent lymphocytes was measured in a 40 hr-^{3}H-proline release test with an effector cell-target cell ratio of 50 : 1. Results are expressed as mean percent cytolysis (%CTL) of triplicate experiments.

Since after BCG no increased cytolysis of nonmelanoma cells was found, it was of interest to know whether this phenomenon was due to a generally lower susceptibility of these cells to lysis. To answer this question, lymphocytes from 31 normal individuals were tested simultaneously on melanoma and nonmelanoma tumor cells. As demonstrated in Fig. 5, the lymphocytes exhibited almost the same mean cytotoxicity against melanoma cells as they did against the nonmelanoma tumor cells. Thus, the nonmelanoma tumor cells were as susceptible to lymphocyte killing as melanoma cells. On the other hand, a considerable portion of the melanoma patients showed a disease-related higher primary cytotoxicity against the melanoma cells that is further increased in a "disease-related fashion" after BCG-administration (Fig.4).

A further aspect of the question concerning the "specificity" of BCG-induced cytotoxic responses is illustrated in Fig. 6, which shows that cytotoxic responses against melanoma target cells occur equally well in autologous as well as allogeneic lymphocyte target cell combinations.

DISCUSSION

The data presented in this paper show that local BCG stimulation induces a systemic short-lasting cytotoxicity response that can be easily monitored in the peripheral blood by use of the ^{3}H-proline release test. This reaction seems to depend on the number of BCG organisms applied. Whether this tumor-directed response measured *in vitro* has a clinical significance is not yet established, although a correlation between the development of cytotoxicity and progression of the tumor disease was recently reported (3). Whether the sharp drop in cyto-

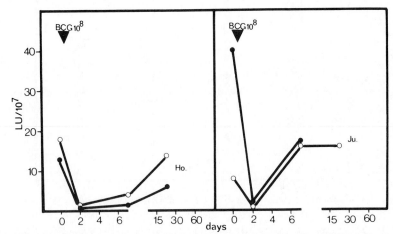

FIG. 6. Short-term effects of high-dose BCG on syngeneic (●——●) and allogeneic (○——○) CMC-systems. Cytotoxicity is expressed as lytic units/10^7 nonadherent lymphocytes (LU/10^7). Target cells: Mel-Ho-88, Mel-JuS-90, and Mel-Ei-78.

toxic activity after high doses of BCG bears any prognostic significance remains to be shown, particularly with regard to a tumor-growth-promoting effect of immunostimulation, observed under certain experimental conditions (4, 6, 9). It is also unclear, whether potentiation of cytotoxicity induced by lower BCG doses parallels an increase in cellular tumoricidal activity *in vivo*. The actual BCG dose leading to such an augmentation of cytotoxicity seems to have a distinct optimum that varies considerably with the individual patient. If these findings can be further substantiated, it is reasonable to postulate that immunostimulation by BCG needs to be individually tailored and monitored. The current lack of individual dosage of BCG in clinical trials may in part explain the failure to prove a general therapeutic efficacy of BCG and may also explain the so far anecdotal beneficial responses in individual patients.

We do not know the underlying mechanism leading to augmentation or decrease of cytotoxicity after BCG treatment. The latter may be due to induction of suppressor cells (7, 8), to generation of blocking factors (11-13), or to changes in the peripheral blood cell composition. Such changes may be brought about by either trapping of cytotoxic cells in the BCG lesion or by mobilization of cells from the central lymphoid organs. With regard to a possible generation of blocking factors, we have tested BCG-dependent alteration of preexisting serum effects on cell-mediated cytotoxicity such as serum-blocking activity or serum-dependent potentation. Simultaneously, the serum titers of antimelanoma antibodies were followed by indirect immunofluorescence in malanoma patients during BCG treatment. So far no characteristic pattern of serum activities could be found that would explain the above-described cytotoxic responses (14).

As to the nature of the effector cells involved, it seems reasonable to conclude that other cells than classical T lymphocytes are instrumental in the cytolytic mechanism. Thus, lack of demonstrable allogeneic restriction, inhibition by Ig aggregates, and resistance of effector cells to heterologous anti-T-cell serum all point to a cytotoxic cell similar to the natural killer cell of the mouse.

To find out whether in addition to the dose of BCG other immunologic parameters influenced the cytotoxic response, we examined the patients' reaction to PPD and DNCB, the local reaction to BCG, and the clinical status of each patient, particularly with regard to an involvement of lymph nodes. Although the number of 18 cases did not yet allow a statistical evaluation, so far no indication for such correlations was found.

Recently it has been reported that BCG can render lymphocytes from healthy donors cytotoxic rather selectively for melanoma cells by mere incubation with the vaccine *in vitro* (15). This observation and our finding of increased cytotoxicity expressed preferentially against melanoma target cells fits into the picture of antigenic similarities between BCG components and the melanoma cell surface (16). This line of reasoning would then favor an induced immune response. On the other hand, spontaneous or natural killer cells do show selective cytotoxicity and preferential lysis of melanoma cells presumably effected by a noninduced mechanism. Possible explanation for the mechanism of BCG-induced boost of cytoxicity may be the modulation of cytolytic effector cells by humoral factors such as interferon as reported by Trinchieri at this symposium.

From the preliminary data presented, one can conclude that continuous assessment of cytolytic activities of circulating lymphocytes during BCG therapy may provide not only clinically meaningful information on the individual patient under immunotherapy, but may also give some insight into the mode of action of BCG.

Acknowledgments

We are indebted to Dr. Jan Fogh, Sloan Kettering Institute, New York, who provided some of the tumor cell lines as indicated. We thank M. Babbel, B. Ermler, and F. Frank for excellent technical assistance and C. Krauss for help with the manuscript.

References

1. Herberman, G. B., and Holden, H. T. *Adv. Cancer Res.* **27**, 305, 1978.
2. Riethmüller, G., Saal, J. G., Rieber, E. P., Ehinger, H., Schnellen, B., and Riethmüller, D. *Transplan. Proc.* **7**, 495, 1975.
3. Saal, J. G., Riethmüller, G., Rieber, E. P., Hadam, M., Ehinger, H., and Schneider, W. *Cancer Immunol. Immunother.* **3**, 27, 1977.
4. Bansal, S. C., and Sjögren, H. O. *Int. J. Cancer* **11**, 162, 1973.
5. Bast, R. C., Zbar, B., Borsos, T., and Rapp, H. J. *N. Engl. J. Med.* **290**, 1413, 1974.
6. Chee, D. O., and Bodurtha, A. J. *Int. J. Cancer* **14**, 137, 1974.
7. Florentin, J., Huchet, R., Bruley-Rosset, M., Halle-Pannenko, O., and Mathé, G. *Cancer Immunol. Immunother.* **1**, 31, 1976.
8. Geffard, M., and Orbach-Arbouys, S. *Cancer Immunol. Immunother.* **1**, 41, 1976.
9. Wepsil, H. T., Harris, S., Sander, J., Alains, J., and Morris, H. *Cancer Res.* **36**, 1950, 1976.
10. Saal, J. G., Rieber, E. P., and Riethmüller, G. *Scand. J. Immunol.* **5**, 455, 1976.
11. Baldwin, R. W., Price, M. R., and Robins, R. A. *Br. J. Cancer* **28**, Suppl. I, 37, 1973.
12. Kamo, J., and Friedman, H. *Adv. Cancer Res.* **25**, 271, 1977.
13. Theofilopoulos, A. N., Andrews, B. S., Krist, M. M., Morton, D. L., and Dixon, F. J. *J. Immunol.* **119**, 657, 1977.
14. Saal, J. G. *et al.* In preparation.
15. Sharma, B., Tubergen, D. G., Minden, P., and Brunda, M. J. *Nature* **267**, 845, 1977.
16. Minden, P., Sharpton, T. R., and McClatchy, J. K. *J. Immunol.* **116**, 1407, 1976.

SUSCEPTIBILITY OF CLONED MELANOMA
TO NATURAL CYTOTOXICITY

I. Korn-Nitschmann, H. H. Peter, E. Krapf, E. Krmpotic,
L. Krmpotic, J. P. Cesarini, and W. Leibold

Cell-mediated cytotoxicity against a variety of tumor target cells has been attributed to several effector cell types including activated macrophages (1), cytotoxic T lymphocytes (CTL) (2), and Fc-receptor bearing null cells (3). Whereas the macrophages express primarily a nonspecific cytostatic activity within the tumor infiltrate (4), CTL with restricted specificity for autologous tumor targets have been identified in the regional lymph nodes (5). Fc-receptor bearing null cells, known to mediate as K cells the antibody-dependent cellular cytotoxicity (ADCC) (6-8) and as NK cells the spontaneous lymphocyte-mediated cytotoxicity (SLMC) (8-10), are predominatly found in peripheral blood (11, 12). Surface marker analysis has revealed that these nonphagocytic K and NK cells are distinct from mature T and B lymphocytes in that they are lacking surface membrane immunoglobulin and fail to form high-affinity E-rosettes (13, 14).

Using allogeneic and autologous melanoma target cell cultures, we have previously established that cytotoxic effector cells in the peripheral blood of healthy donors and melanoma patients belong to the group of K and NK effector cells (8, 12, 15). Whereas the specificity of the K cell reaction (ADCC) is determined by the target-sensitizing antibody, the NK cell effect (SLMC) might involve different effector molecules of unknown specificity such as natural antibodies (16), lymphotoxin (17), or interferon (18). Moreover SLMC, unlike ADCC, seems to be crucially dependent on the type of target cell employed; this facet of the reaction has so far received comparatively little attention.

99

In the present study we have examined the target cell dependence of an SLMC reaction involving allogeneic normal human effector cells and a panel of 14 target cell cultures from a melanoma patient (ST). It was found that clonal melanoma subcultures vary greatly in their susceptibility to lysis, not only on a given day but also when tested sequentially. Fibroblasts and an EBV-transformed B-lymphoblastoid cell line (LCL) from the same patient were resistant to the NK effect. Conditions possibly related to the observed changes of target cell sensitivity were examined and discussed in view of the difficulties encountered in using SLMC assays for sequential testing of patients.

MATERIAL AND METHODS

Target Cell Lines

A melanoma cell culture (ST) was established from an inguinal lymphnode metastasis of a 34-yr old woman. Blood group, HLA type, clinical history and treatment of the patient are summarized in Table 1. A fibroblast culture (ST-Fib) and EBV-transformed B-LCL (ST-EBV) were also established from the same patient. The melanoma origin of the established tumor cell culture was proven by the histology of the tumor grown in nude mice, the electronmicroscopic demonstration of intracytoplasmic melanosomes (Fig. 1) and by the detection of 5-S-Cysteinyl-Dopa in the culture medium (19). Between the ninth and the 14th passage, clonal subcultures were established by seeding limiting dilutions of melanoma cells (2 to 20 cells per well) in flat-bottom microculture plates (Falcon No. 3040; COSTAR, Cambridge No. 3524). Sterile glass or plastic coverslips (Eurolalab, Thermanox No. 5414) were placed on the bottom of each well before addition of the tumor cells. The plates were monitored daily under an inverted micro-

TABLE 1. Origin of the Melanoma Target Cell Lines

34-year old woman (ST)

Blood group: A1B, Rh pos; *HLA* : A2, AW23, BW15, BW40, CW2, CW3

Clinical history
    November 1975 excision of primary melanoms, left foot.
    November 1976 node recurrence left groin, radical surgery.
    November 1976-April 1977 six courses of DTIC.
    April 1977 lymphnode recurrence left groin, surgery.
    May-July 1977 CCNU; August 1977 dissemination; September 1977 death.

Establishment of cell lines:
    November 1976, first melanoma culture (ST), served for cloning.
    April 1977 fibroblast culture and second melanoma culture, both lost.
    September 1977, EBV transformed B-LCL.

scope. Wells with one or two single adherent cells were marked and checked for clonal proliferation. As soon as clones developed, the coverslips were removed, eventually cut in half to separate two clones, and the isolated clones were placed separately into tissue culture flasks (Falcon No. 3013). After 8 to 12 weeks, the clonal subcultures had sufficiently expanded and could be used as target cells. Ten clonal subcultures, designated B, C, D, E, G, H, J, K, L, and M were raised from the initial melanoma culture ST. An eleventh subculture derived from the same tumor was grown for four weeks in a nude mouse. Together with the fibroblasts (ST-Fib) and the B-LCL (ST-EBV) cells, we had 14 different cultures from the same patient.

The myelogenous suspension line K 562 (20) and the B-LCL Raji cells derived from an African Burkitt lymphoma biopsy (21) served as controls. All cell cultures as well as the SLMC assays were performed with Eagle's minimal essential medium (MEM, Auto-Pow, Flow Laboratory, 53 Bonn, GFR), supplemented with 10% heat-inactivated fetal calf serum (FCS), pyruvate, glutamine, nonessential aminoacids, vitamins, and antibiotics (15). All cultures were routinely checked for mycoplasma contamination at the Institut für Microbiologie, Med. Hochschule Hannover, and found to be sterile.

Growth Curves

Growth curves were established for the melanoma culture ST and the clonal subcultures by explanting 5×10^3 cells/well in flat-bottom culture plates (COSTAR, Cambridge, Cat. No. 3524) equipped with round coverslips. At 24, 48, 72, and 96 hr, five wells of each culture were pulsed for 6 hr with 1 μCi of ^{3}H-thymidine (^{3}H-TdR). The coverslips were then removed, washed, and counted for radioactivity in a Packard Scintillation Counter using Ista-Gel, Packard Co. Coverslips without cells were treated identically and served as controls. The mean ^{3}H-TdR uptake ±1 standard error (SE) was logarithmically plotted against culture time, and doubling times were calculated.

Karyotyping

Chromosomal studies of the original melanoma culture ST and the 10 clonal subcultures were performed using the method of Moorhead et al. (22). The conventional staining technique and the Giemsa banding technique (23) were applied. From each cell line 12 to 48 metaphases were photographed. The chromosomes were counted on 18 x 24 cm enlarged prints of the 35 mm negatives. Karyotypes were prepared according to the Paris Conference-nomenclature (24).

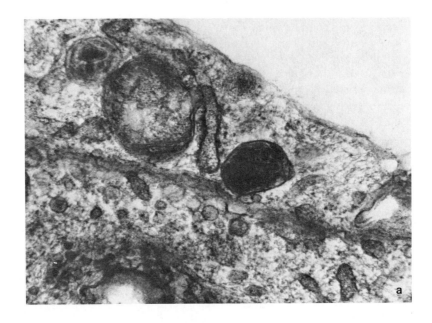

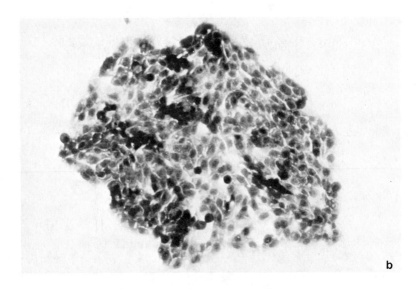

FIG. 1.   (a) Electronmicroscopic demonstration of melanosomes; (b) clonal subculture of melanoma ST; (c) and (d) melanoma ST grown in BALB/c.*nu/nu*.

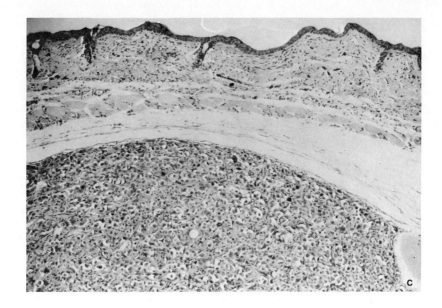

FIG. 1(c).

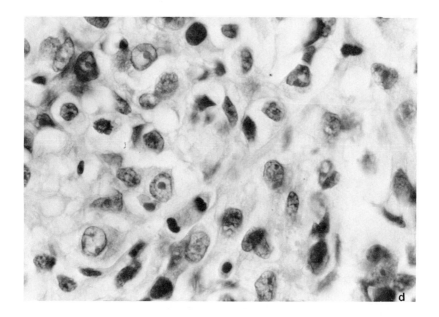

FIG. 1(d).

Effector Cells

Peripheral blood lymphoid cells from three healthy donors (two males, 27 and 35 yr old, one female, 25 yr old) were isolated three times monthly at intervals by Ficoll-Hypaque density gradient centrifugation (25). From a fourth donor (30-yr old man), cryopreserved lymphocytes were compared on three occasions to his fresh lymphocytes. In one of these experiments, fresh lymphocytes depleted of mononuclear phagocytes by iron-magnetism and plastic adherence (8) were tested in parallel.

^{51}Cr Release Assays

SLMC assays were performed in round bottom microculture plates (Linbro, Titertek, LS Labor-Service, 8000 Munchen, GFR). 10^6 target cells from 5-day old subcultures were labeled with 100 μCi of $Na_2\,^{51}CrO_4$ (specific activity 200 to 400 μCi/μg; EIR, 53/3 Würenlingen, Switzerland) according to previously described

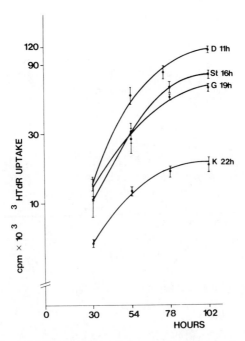

FIG. 2. Growth curves of melanoma line ST and three clonal subcultures (D, G, K) measured by 3H-TdR incorporation.

protocols (8). 10^4 labeled and washed targets were distributed in 0.1 ml aliquots per well to the microplates. Triplicates were set up for the lymphocyte to target (L/T) ratios of 3:1, 10:1, 33:1, and 100:1. The microplates were spun for 1 min at 200 x g and then incubated for 12 hr at $37°C$ in an atmosphere of 5% CO_2 in air. Using the Skatron disposable supernate collecting device (DSC, Titertek, Flow Laboratory, 53 Bonn, GFR), it became technically possible to test on one day effector lymphocytes from up to four donors against 16 to 20 different target cells. Results were expressed either as percent specific ^{51}Cr release = (cpm test–cpm background):(cpm input–cpm background) x 100 or as arbitrary lytic units per 10^6 mononuclear cells, based on 15% specific ^{51}Cr release (2).

Since the target cell labeling and the subsequent SLMC assays were performed on suspended monolayer cells, we investigated in a first experiment the influence of various cell detaching procedures on target cell susceptibility to lysis. Three-minute treatment with 0.02% EDTA plus vigorous shaking was compared to 3- or 15-min treatment with 0.02% EDTA + 0.05% trypsin (Grand Island Biological, Cat. No. 530). All subsequent experiments were performed with target cells suspended by a 3-min treatment with EDTA-trypsin prior to isotope labeling.

RESULTS

Morphology and Growth Curves of Clonal Melanoma Subcultures

Using standardized culture conditions, clonal subcultures from an explanted human melanoma exhibited considerable differences in their doubling times as measured by the incorporation of ^{3}H-TdR. Figure 2 depicts four typical growth curves. The differences in proliferation were also reflected by shorter or longer culture periods necessary to obtain confluent monolayers. A particularly slow growth rate was observed with subculture K, whereas D and J showed the fastest proliferation. Morphological differences between the clonal subcultures were less striking; all cultures were composed of typical epithelioid cells (Fig. 1) with more or less pronounced dentritic plasma membrane protrusions.

Karyotyping of Clonal Melanoma Subcultures

A preliminary chromosomal analysis of the various clonal subcultures showed a remarkable cellular heterogeneity even within a given clone. Besides a high variability in the chromosome number indicating aneuploid cells, the clonal subcultures seem to express certain marker chromosomes in a quantitatively different fashion. Whereas an abnormal chromosome, designated M2 (Fig. 3), was found in most of the examined mitoses of all clonal subcultures, three bicentric marker chromosomes—B1, B2, B3—and a chromosomal fragment F (Fig. 4) showed different clonal distributions (Table 2). A cautious interpretation of this preliminary chromosomal analysis would suggest that our clonal subcultures are not true clones derived from a polyclonal tumor, but rather represent different clonal evolutions of one initial malignant cell clone.

Spontaneous Lymphocyte-Mediated Cytotoxicity (SLMC) Assays

Effect of target cell pretreatment with EDTA and trypsin.

A first experiment was designed to test the influence of EDTA and EDTA-trypsin pretreatment on target cell susceptibility to SLMC. Cells from six clonal subcultures were removed from the tissue culture flasks either by 3-min exposure to 3 ml of 0.02% EDTA, accompanied by vigorous shaking, or by 3- or 15-min

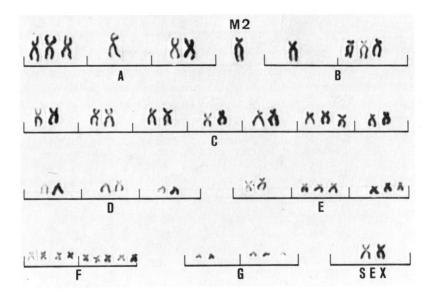

FIG. 3. Karyotype of an aneuploid melanoma cell (ST) with marker chromosone M2 in the second chromosome group.

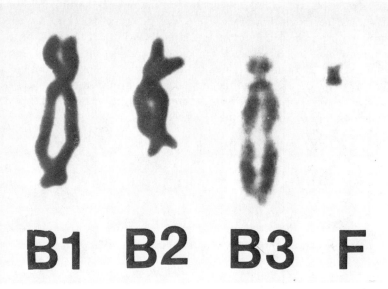

FIG. 4. Three types of bicentric marker chromosomes and chromosomal fragment F. The bicentric chromosomes showed varying distributions among the different clonal subcultures, whereas F and M2 were present in nearly all examined mitoses of the different melanoma clones.

treatment with 3 ml of 0.02% EDTA + 0.05% trypsin. All cells were washed, labeled, and used as targets. Isotope uptake and specific release after incubation with normal effector cells are illustrated in Fig. 5. Compared to EDTA treated targets, it can be seen that a 3-min exposure of melanoma cells to EDTA-trypsin resulted in a significantly increased isotope uptake, an unchanged background release at 12 hr, and an increased specific release. Subsequently, this treatment was chosen to suspend the target cells prior to labeling. The K562 suspension line was not significantly affected by any of the three treatments.

Clonal variations in target cell susceptibility to lysis.

Effector lymphocytes were tested against a panel of normal and malignant cells from the same donor. Figure 6 illustrates three SLMC assays performed at monthly intervals with effector lymphocytes from three healthy donors and target cells from the same clonal melanoma subcultures. Whereas the normal fibroblasts and the B-LCL obtained from the same patient (ST) were resistant to lysis, the clonal melanoma subcultures exhibited great differences in sensitivity ranging from resistance to high susceptibility to lysis. The target cell sensitivity was un-

TABLE 2. Chromosomal Analysis of Clonal Melanoma Subcultures Derived from an Inguinal Lymphnode Metastasis of Patient ST

Clonal subcultures	No. of Mitoses evaluated	No. of Chromosomes (range)	Chromosomal markers					Chromosomal fragments
			M2	B1	B2	B3	F	
ST14[a]	35	59–129	+++	+	++	+	+	
B	25	36–215	++	+	++		++	
C	50	37–216	++		++		+	±
D	20	74–342	+		++		+	+
J	25	48–409	+++	+	+		+	
E	25	53–225	++	+	+	+	+	
G	20	52–332	++	+			+	+
H	25	46–146	+			++	+	+
K	12	34–161	+	+	+	+	+	±
L	23	58–279	+++	+	+	+	++	
M	24	55–202	++		+		+	+

[a] Initial melanoma culture in passage 14. Single cell clones B, C, D, ... M were established from this culture between passage 9-14.

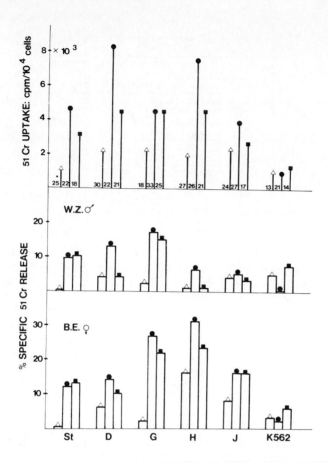

FIG. 5.   Effect of target cell pretreatment with EDTA (0.02%) or EDTA (0.02%) + trypsin (0.05%) for 3 or 15 min on isotope uptake and specific ^{51}Cr release. Six targets and two effectors were tested. The background release (*) did not change significantly in the differently pretreated target cell populations.

stable when tested sequentially over a three-month period. Besides regularly resistant (E, M) or susceptible clones (G, D), there were others in which susceptibility rose (NM) or rose and fell (ST, B, C, H, J, K, L). Compared to some of the clonal subcultures, the initial uncloned melanoma culture ST proved to be less variable in its target cell properties. A development from high initial susceptibility to lysis to almost complete resistance was observed with the K562 control line. So far, this line remained resistant to effector cells from ten other donors. There is no evidence that our K562 line has been contaminated by other cells. It is still EBNA negative and reacts with heteroantisera raised against human tissue culture cell lines. It does not react with an antiserum against the murine L1210 line, which is by morphological criteria the only similar suspension line maintained in our laboratory.

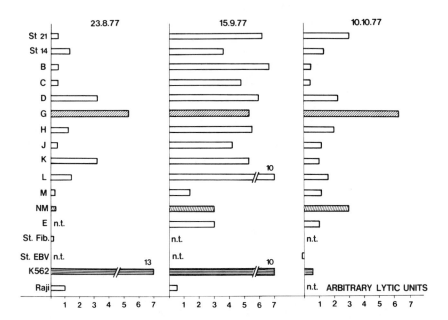

FIG. 6. Three sequential SLMC assays performed with lymphocytes from an individual donor (Hans) and a panel of target cells consisting of the melanoma line ST grown *in vitro* or in a nude mouse (NM), 10 clonal melanoma subcultures (B to M), fibroblasts (ST-Fib), B-LCL (ST-EBV), and the suspension lines K562 and Raji.

To test whether target cell variations may occur within the same passage of a clonal subculture, stored and fresh effector lymphocytes from the same individual were tested on targets obtained from nonconfluent (2 days), confluent (4 days), and crowded (6 days) clonal melanoma subcultures (D, L, M). The results of this experiment are summarized in Fig. 7. Whereas cryopreserved effector cells were in general slightly more cytotoxic than fresh lymphocytes, both effectors produced remarkably parallel effects on targets from 2-, 4- or 6-day old cultures. The targets from 4-day old cultures proved to be most resistant, whereas cells from 6-day old cultures showed the highest susceptibility to lysis. These results strongly suggest that day-to-day variations of the target cell sensitivity have an important impact on the outcome of an SLMC assay. The higher cytotoxic efficiency of cryopreserved over fresh effector cells is likely to be due to a relative enrichment of NK cells during cyropreservation; mononuclear phagocytes and granulocytes known to lack NK activity against melanoma cells, are lost (8, 15). In fact, removal of the mononuclear phagocytes raised the SLMC activity of fresh effector cells to the level of the cryopreserved lymphocytes (Fig. 7).

In addition to changes in target cell sensitivity, unexplained day-to-day variations may also occur in the effector cell population. This is suggested by the sequential testing of lymphocytes from three healthy donors against the same target cell panel (Fig. 8).

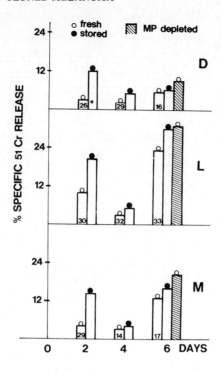

FIG. 7. Three sequential SLMC assays performed with fresh, and cryopreserved effector lymphocytes from the same donor and target cells obtained from nonconfluent (2 days), confluent (4 days) and crowded (6 days) melanoma cultures (clones D, I, M). On day 6 also fresh lymphocytes depleted of mononuclear phagotypes (shaded columns) were tested.

DISCUSSION

The main issue in SLMC is the characterization of effector cells (3, 8, 9, 10). Recently, analysis of the specificity of the SLMC reaction has received greater attention, but little is known about the target cell structures necessary to elicit SLMC. None of the proposed effector mechanisms for SLMC—natural antibodies (16), lymphotoxin (17), or interferon (18)—have explained the selective killing of particular target cells (26) or day-to-day variations in target cell sensitivity. Petranyi et al. (27) and Santoli et al. (38) presented evidence that SLMC may be controlled by genes close to the major histocompatibility complex; it appears, however, that fluctuations in target cell susceptibility to lysis can greatly influence the outcome of SLMC reactions and thus mask a possible genetic control of the effector mechanism.

Differences in target cell susceptibility to NK lysis have been reported by several authors. Jondal et al. (9) differentiated between fast and slowly lysable

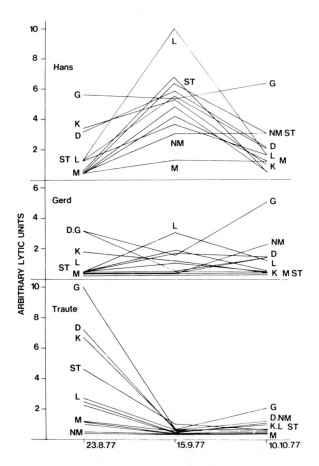

FIG. 8. Sequential SLMC assays were run at monthly intervals with effector lymphocytes from three donors (Hans, Gerd, Trude) against a panel of clonal melanoma target cells. Note clonal target cell variations in the assays performed with effectors from Hans and Gerd and the unexplained drop of SLMC activity with lymphocytes from Trude.

cells, whereas Santoli *et al.* (28) distinguished target cell lines of high, intermediate, or low lytic susceptibility. Ono *et al.* (29) reported a higher sensitivity of human T-LCL as compared to B-LCL. Experiments performed by Leibold *et al.* confirmed and extended these findings (30, 31). With regard to variations in lytic suscepti-bility of human melanoma cells, DeVries *et al.* (32) and Mukherji *et al.* (33) found target cells from short-term melanoma cultures less susceptible to lysis than long-term melanoma lines. In our personal experience with the IGR3 melanoma line (8, and unpublished observations), this finding did not apply to cultures beyond the tenth passage. In fact, unexplainable variations in the lytic susceptibility of IGR3 cells in different passages prompted us to undertake this study. Our ap-proach was to use cloned and uncloned melanoma cells, fibroblasts, and B-LCL from one patient and to expose all cells simultaneously to effector cells from a normal donor. The results show that the melanoma cells were clearly more susceptible to SLMC than B-LCL and fibroblasts. Tumor cells, fibroblasts, and B-LCL have all been shown to express HLA-A, B, and C locus antigens in variable quantities (34). Whereas the fibroblasts and tumor cells tend to lose their antigens with time in culture (34), B-LCL maintain high-absorption capacities for HLA antibodies (35). Therefore, our data would argue against HLA-A, B, and C locus antigens being the relevant target cell determinants for SLMC. The involvement of D or DRw antigens is also unlikely on the basis of the presented results, since neither the antigen-rich B-LCL nor the fibroblasts, known to express little or no D or DRw antigens, were lysed.

Another explanation for the increased lysis of melanoma cells might be the possible presence of tumor-associated antigens similar to those described in murine target cell lines susceptible to NK lysis (36).

Differences in susceptibility to lysis were not only observed among clonal melanoma subcultures but also within several clones when tested sequentially against the same effector cells. This heterogeneity of the clonal melanoma subcultures with respect to NK lysis may explain the similar but generally less pronounced fluctuations observed in the uncloned culture.

The variations in target cell susceptibility to SLMC occurred irrespective of the chromosomal constitution; this suggested that the different melanoma subcultures represent clonal evolutions of only one malignant cell clone. As to marker chromosomes, there was no obvious relationship to lysis in SLMC assays. Far more important for alterations or target cell sensitivity appeared to be the culture conditions and the pretreatment of the target cells prior to labeling. Cells from non-confluent, confluent, and crowded cultures exhibited different degrees of lytic susceptibility. Due to different growth rates, the various clonal subcultures reached confluence after different culture periods; since only cells from 5-day old cultures were used as targets, this may explain some of the observed heterogeneity. Important changes of target cell susceptibility to SLMC could also be introduced by the detachment procedures used to suspend monolayer target cells prior to the isotope labeling. With the melanoma cultures examined in this study, a 3-min treatment with EDTA-trypsin significantly increased the isotope uptake as well as the subsequent specific target cell lysis. This phenomenon may not be relevant to all target cells, since the K562 suspension line was not significantly affected by this treatment.

The myelogenous suspension line K562 was originally established by Lozzio and Lozzio (20) and later employed by Klein et al. (37) as a highly susceptible target cell line for SLMC. During a 4-month test period, we observed a gradual loss of lytic susceptibility of this line; this may be explained by the clonal evolution and expansion of a lysis-resistant cell clone or by a reduced expression of SLMC eliciting membrane antigens. The switch from a susceptible to a resistant target cell, or vice versa, may also be brought about by metabolic changes and membrane alterations related to the cell cycle, as has been shown for the complement-dependent immune cytolysis (38, 39). Furthermore, the turnover of certain membrane components (e.g., enzymes), the cellular repair mechanism, changes in the electrical membrane potential, or susceptibility to viral infections may have profound effects on the lytic susceptibility of target cells used in SLMC reaction. Another approach to explain the variations in lytic susceptibility has been reported by Trinchieri (18). He suggested that interferon being produced during the SLMC reaction can modulate both the effector cell activity and the target cell sensitivity.

The data presented indicate that high, intermediate, or low susceptibility to NK lysis are relative properties of target cells, which may change with time in culture and can, therefore, influence an objective estimation of effector cell activities. This observation raises particular problems for all clinically oriented SLMC studies involving sequential assays with effector cells from a given donor. The use of a large target cell panel in sequential SLMC assays may circumvent part of the problems related to the target cell variability (40).

Acknowledgment

The EBV-transformed B-LCL line was established and kindly given to us by Ms. Hagedorn and Prof. V. Viehl, Abteilung fur Hämatologie der Medizinische Hochschule, Hannover. The excellent technical assistance of Ms. Hannelore Dröge and Ms. Anette Servin is gratefully acknowledged.

This work was supported by DFG Sachbeihilfe Pe 151/5 and SFB 54/C5 + F2.

References

1. Hibbs, J. B. *In* "The Macrophage in Neoplasia" (M. A. Fink, ed.). p. 83. Academic Press, New York, 1976.
2. Cerottini, J. C., and Brunner, K. T. *Adv. Immunol.* 18, 67, 1974.
3. Bonnard, G. D., and West, W. H. Cell-mediated cytotoxicity in human. *In* "Immunodiagnosis of Cancer" (R. B. Herberman and K. R. McIntire, eds.). Marcel Dekker, New York, in press.
4. Eccles, S. A. Macrophages and cancer. *In* "Immunological Aspects of Cancer" (J. E. Castro, ed.). University Park Press, Baltimore, 1977.
5. Vanky, F. Human tumor-lymphocyte interactions *in vitro*. *In* "Recent Trends in Immunodiagnosis and Immunotherapy of Malignant Tumors" (C. Herfahrth, H. D. Flad, eds.). Springer Verlag Heidelberg, New York, in press.
6. MacLennan, I, C. M. *Transpl. Rev.* 13, 67, 1972.
7. Perlmann, P., Perlmann, H., and Wigzell, H. *Transpl. Rev.* 13, 91, 1972.
8. Peter, H. H., Pavie-Fischer, J., Fridman, W. H., Aubert, C., Cesarini, J. P., Roubin, R., and Kourilsky, F. M. *J. Immunol.* 115, 539, 1975.
9. Jondal, M., and Pross, H. F. *Int. J. Cancer* 15, 596, 1975.
10. Pross, H. F., and Jondal, M. *Clin. Exp. Immunol.* 21, 226, 1975.
11. Vose, B. M., Vanky, F., Argov, S., and Klein, E. *Eur. J. Immunol.* 7, 753, 1977.

12. Kalden, J. R., Peter, H. H. Roubin, R., and Cesarini, J. P., *Eur. J. Immunol.* 7, 537, 1977.
13. Chess, L., MacDermott, R. P., Sondel, P. M., and Schlossman, S. F. *Progr. Immunol.* *II* 3, 125, 1974.
14. West, W. H., Cannon, G. B., Kay, H. D., Bonnard, G. D., and Herberman, R. B., *J. Immunol.* 118, 355, 1977.
15. Peter, H. H., Knoop, F., and Kalden, J. R., *Z. Immun. Forsch.* 151, 263, 1976.
16. Troye, M., Perlmann, P., Pape, G. R., Spiegelberg, H. L., Naslund, I., and Gidlöf, A. *J. Immunol.* 119, 1061, 1977.
17. Peter, H. H., Eife, R. R. and Kalden, J. R. *J. Immunol.* 116, 342, 1976.
18. Trinchieri, G. This volume.
19. Prota, G., Rorsman, H., and Rosegren, G. *Experientia* 32, 970, 1976.
20. Lozzio, C. B. and Lozzio, B. B. *Blood* 45, 321, 1975.
21. Epstein, M. A., Achong, B. G., Barr, Y. M., Zayac, B., Henle, G., and Henle, W. *J. Natl. Cancer Inst.* 37, 547, 1966.
22. Moorehead, P. S., Nowell, P. C., Mellman, W. C., Battips, D. M., and Hugerford, D. A. *Exp. Cell Res.* 20, 613, 1960.
23. Seabright, M. *Lancet* 2, 115, 1971.
24. Paris Conference on Standardization in Human Cytogenetics (1971) Birth defects. Original Article Series **VIII**, 7 (1972). The National Foundation, New York.
25. Böyum, A. *Scand. J. Clin. Lab. Invest. Suppl. 97* 21, 7, 1968.
26. Takasugi, M., Akira, D., Takasugi, J., and Mickey, M. R. *J. Natl. Cancer Inst.* 59, 69, 1977.
27. Petranyi, G., Benczur, M., Onody, C. E., and Hollan, S. R. *Lancet,* 1, 736, 1974.
28. Santoli, D., Trinchieri, G., Zmizewski, C. M., and Koprowski, H. *J. Immunol.* 117, 765, 1976.
29. Ono, A., Amos, D. B., and Koren, H. S. *Nature* 266, 546 1977.
30. Leibold, W., Peter, H. H., and Gatti, R. A. *In* "Mechanisms in Regulation of the Immune Response in Cancer" (A. Mitchison, M. Landy, eds.). Academic Press, New York, 1979.
31. Leibold, W., Janotte, G., and Peter, H. H. *Z. Immun. Forsch.* 153, 328, 1977 (Abstract).
32. DeVries, J. E., Cornain, S., and Rümke, P. *Int. J. Cancer* 14, 427, 1974.
33. Mukherji, B., Vassos, D., Flowers, A., Binder, S. C., and Nathanson, L. *Int. J. Cancer* 16, 971, 1975.
34. Sasportes, M., Dehay, C., and Fellous, M. *Nature* 233, 332, 1971.
35. Gatti, R. A., Svedmyr, E. A. J., Leibold, W., and Wigzell, H. *Cell. Immunol.* 15, 432, 1975.
36. Herberman, R. B., Nunn, M. E., and Lavrin, D. H. *Int. J. Cancer* 16, 216, 1975.
37. Klein, E., Ben-Bassat, H., Neumann, H., Ralph, P., Zeuthen, I., Polliack, A., and Vanky, F. *Int. J. Cancer* 18, 421, 1976.
38. Lerner, R. A., Oldstone, M. B. A., and Cooper, N. R. *Proc. Natl. Acad. Sci. USA* 68, 2584, 1971.
39. Pellegrino, M. A., Ferrone, S., Cooper, N. R., Dierich, M. P., and Reisfeld, R. A. *J. Exp. Med.* 140, 578, 1974.
40. Herberman, R. B., and Oldham, R. K. *J. Natl. Cancer Inst.* 55, 749, 1975.

ANTIGEN RECOGNITION BY CYTOTOXIC
T LYMPHOCYTES

Kirsten Fischer-Lindahl

Cytotoxic T lymphocytes (CTL) are uniquely suited for a study of antigen recognition by T cells, since, once generated, they are autonomous in their actions (1), and one can even watch in the microscope how a single CTL kills its target (2). The cytotoxic reaction is therefore not complicated by the specificity requirements of an unknown number of cell interactions as are T-cell help for antibody formation, T-suppressor cell activity or delayed hypersensitivity (3).

T-cell-mediated lysis is specific (4). The specificity is mediated by surface receptors on the CTL which allows it to bind to the target cell. Once appropriate juxtaposition of killer and target cell has been attained, killing proceeds in a non-specific manner (5, 6). Mere binding of killer to target cell is insufficient for killing, since *A* anti-*B* CTL will bind to and kill *B* anti-*C* CTL, but the former are not in turn killed by the *B* anti-*C* CTL (7). The receptors of the CTL can be demonstrated by the specific binding of CTL to target cell antigens (8-10). The capacity of CTL to bind antigen and kill is sensitive to proteolytic enzymes but is spontaneously regenerated by purified T cells at $37°C$ (8, 11, 12). Available evidence indicates that CTL synthesize their own receptors, as precursors of cytotoxic lymphocytes are already predetermined in their specificity (13), and that B cells are not required for the generation of CTL (14). *In vitro* generated CTL have no Fc receptors (15), and Fc receptors are not required for cytolytic activity (16).

TABLE 1. Xenogeneic CTL are Specific for H-2K and H-2D[a]

Stimulator	Origin of H-2 region								Percent ^{51}Cr-release (mean ± SD) from targets			
	K	A	B	J	E	C	S	D	B10.HTT	AQR	B10.DA	A.SW (H-2^s)
B10.HTT	s	s	s	s	k	k	k	d	21.3 ± 2.6 (H-2)	15.0 ± 1.7 (D)	5.8 ± 7.2 (—)	3.9 ± 0.7 (K, I)
AQR	q	k	k	k	k	d	d	d	18.4 ± 1.7 (D)	29.8 ± 6.4 (H-2)	13.0 ± 0.5 (K)	-0.1 ± 1.8 (—)
B10.DA	a	q	q	q	q	q	q	s	1.8 ± 0.2 (—)	7.2 ± 3.6 (K)	19.6 ± 0.8 (H-2)	7.1 ± 2.2 (D)
B10.S	s	s	s	s	s	s	s	s	11.9 ± 0.6 (K,I)	2.6 ± 0.3 (—)	18.3 ± 3.3 (D)	19.6 ± 0.8 (H-2)

[a] Peripheral blood lymphocytes from one human donor were sensitized *in vitro* to mitomycin-C-treated mouse spleen cells of the strains shown and after six days of culture their cytotoxicity was assayed in triplicate at 140 effectors per one LPS-stimulated target cell. Spontaneous release ranged from 13 to 20% of total release. In parentheses are indicated *H-2* regions shared between stimulator and target; B10.HTT, B10.DA, and B10.S are identical except for *H-2*, and AQR shares a considerable part of their non-*H-2* genetic material (19).

CTL INTERACT WITH H-2K AND H-2D

The fundamental feature of CTL specificity is that only mouse target cells that share the H-2K or H-2D region with the stimulating cells are lysed (17). This is so even when one studies responding cells of another species that should be able to perceive differences of many other surface antigens (18; Table 1). This is also the case when the responding and stimulating cells have the same H-2 haplotype and differ only in some minor histocompatibility antigens (20, 21; see Fig. 1), or because of viral infection (22) or chemical modification of the stimulating cells (23). Target cells not expressing H-2 are generally not lysed (24-26). There are very few exceptions to this rule. T-cell-mediated cytotoxicity specific for products of the I region can be generated, although not very efficiently, and proceeds independently of the H-2K and D specificities on the target cell (27, 28). CTL specific for target cells carrying the F9 antigen but no H-2 antigens have also been demonstrated (H. Wagner, this volume). It may be significant that the antigens in both of these cases are controlled by the same chromosome as H-2K and D (29). Finally, highly active CTL generated in an allogeneic combination can be shown to kill not only the stimulator but also some third party allogeneic target cell with different H-2 antigens—this appears to be a specific crossreactivity of the H-2 antigens as seen by the CTL (30).

It has often been discussed whether the target molecules for CTLs are indeed the serologically defined H-2K and D molecules or some other cell surface antigens encoded by the same regions of the H-2 complex (see ref. 30 for review). It was shown very early that alloantisera against the H-2 antigens of the target cells block cytolysis (31, 32) and this has been taken as evidence that the serologically defined H-2K and D molecules are involved in the killer-target cell interactions. Such a conclusion is, however, entirely dependent on the purity and specificity of the antiserum used. When monoclonal anti-H-2K^k antibodies of single specificity recently became available from myeloma cell fusions (33), it became important to repeat these inhibition studies (34). Three different antibodies were used. Their reactivity with a panel of H-2 haplotypes suggested that they recognized H-2 specificites 5, 11, and 25, respectively. All three antibodies reacted similarly; for simplicity, only the results obtained with the one that had the median inhibitory capacity are shown (Fig. 1). The monoclonal antibodies block allogeneic cytotoxicity directed against H-2K^k but not that directed against the D region on the same target cell (panels a and d). The antibodies do not interfere with the cytotoxic activity of effector cells carrying H-2K^k [(b x k)F$_1$ responders sensitized to BALB/c (H-2^d) or to B6 (H-2^b) male cells] . The monoclonal antibodies also block H-2 restricted cytolysis specific for minor histocompatibility antigens controlled by the B10 non-H-2 background (34) or by the Y chromosome (Fig. 1, panels b and e). Again, blocking of lysis is specific: only H-Y specific CTL restricted by the D region on the same target cell, are inhibited. These data show that the killer-target cell interaction, whether allogeneic or H-2 restricted, proceeds via the same cell surface molecules.

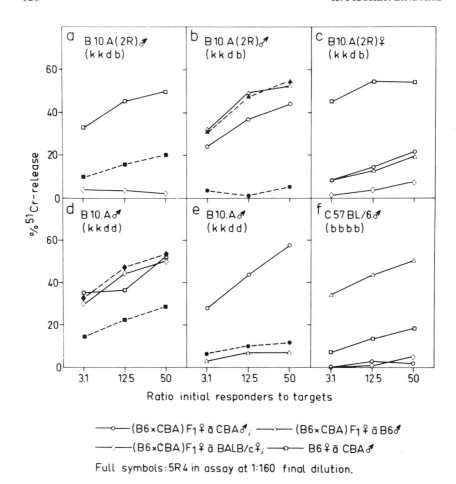

FIG. 1. Inhibition of cytotoxicity by monoclonal BALB/c α-CBA antibody from clone 5R4, which reacts like α-H-2.11. For clarity, the allogeneic cytotoxicity against H-2K^k (–□–) and against H-2D^d (–◇–) are presented in panels a and d, while the H-Y specific cytotoxicity restricted by H-2K^k (–○–) and H-2D^b (–△–) are presented in panels b and e, although all assays were done on the same target cells. Targets, indicated in each panel together with their H-2 haplotypes, were LPS-stimulated spleen cells, and spontaneous release ranged from 23 to 30% of total release. Effector cells were primed *in vivo* with 2×10^7 syngeneic male cells and restimulated *in vitro* (35). The target cells were preincubated with antibodies for ½ hr at 37°C, then effector cells were added and the assays were further incubated 3½ hr at 37°C in the presence of antibodies.

SPECIFICITY OF ALLOANTIGEN-ACTIVATED CTL

CTL activated in a mixed lymphocyte culture where the number of responding cells were nonlimiting, or *in vivo*, are very heterogeneous as evidenced by their cross-reactivity patterns (30, 36). It is important to be able to study the antigen recognition patterns of a single CTL or a clone of CTL derived from a single precursor and therefore presumably with identical specificities.

A limiting dilution-mixed lymphocyte culture system was developed in which it was possible to generate a clone of active CTL from a single killer cell precursor (described in detail in ref. 37). Mixed lymphocyte cultures were set up in V-bottom microtiter plates with one million irradiated stimulator spleen cells in each well and very low numbers of responding cells (in an allogeneic combination, concentrations ranging from 5×10^3 down to 3×10^2/well). To get good responses, it was necessary to use 50% conditioned medium that was derived from a two-day old bulk mixed lymphocyte culture. After six days of culture, ^{51}Cr-labeled target cells were added to each well, and one could determine the frequency of wells with cytotoxic activity at each concentration of responding cells. The Poisson distribution then gives the frequency of the limiting unit from a semilog plot of frequency of negative wells versus number of responding cells. As discussed in detail elsewhere (37), we believe that the limiting factor in our system was the cytotoxic T-lymphocyte precursor (CTL.P), and consequently we determined CTL.P frequencies for a variety of antigens (Table 2). Precursors responsive to an allogeneic MHC stimulus are as frequent as 1 per 800 lymph node cells. If one takes into account that only Ly-2^+, 3^+ T lymphocytes, i.e., 5-10% of peripheral T lymphocytes, have the potential to develop into alloreactive CTL (38), one finds that at least 1-3% of the potential CTL.P are reactive to a single allogeneic MHC haplotype. Similar conclusions have recently been reached by several other groups (6, 39, 40). Less frequent are CTL.P reactive to xenogeneic stimulus or to TNP-modified syngeneic cells (Table 2).

The number of CTL generated in a positive culture is sufficient to allow one to split each culture on the day of assay and compare its cytotoxic activity on two or perhaps even more target cells. Thus we found that the majority of CTL.P activated by one allogenic stimulus would generate CTL toxic only to the specific stimulator, but a few of the clones would also kill allogeneic target cells of a third haplotype (37). With a whole panel of different target cells, this method can be used to estimate the minimal number of different specificities among CTL.P reactive to one haplotype.

As discussed, the interaction between killer and target cell proceeds via the same H-2 molecules, whether it is an allogeneic reaction (where the target cell

TABLE 2. Frequencies of CTL.P Among Lymph Node Cells

Combination	Number of cells with 1 CTL.P		
	Median	Range	n^a
C57BL/6 ā DBA/2	816 (2.3%)^b	677-935	5
C3H ā DBA/2	1398 (1.4%)	893-2264	12
H-2^k or H-2^d ā C57BL/6	3590 (0.5%)	1438-11,527	10
Mouse ā BN rat	5380	4045-11,045	4
C57BL/6 or DBA/2 ā TNP-self	15,947	8675-39,103	5

^{a} Number of experiments.

^{b} Estimated frequency among Ly-2+, 3+ T cells.

presents an H-2 molecule different from those of the killer) or it is an H-2 restricted cytolytic reaction (where killer and target possess the same H-2 antigens).
It is important to know whether the two reactions might be mediated by the
same cells. Cantor and Boyse (38) presented evidence that Ly $1^+, 2^+, 3^+$ T cells
are required for generation of antiself TNP CTL, but not for generation of alloreactive CTL, suggesting that there might be two different populations of killer
cells. However, the difference could be due to a requirement for a helper during
the activation against TNP-self (41), since the mature CTL in both cases are Ly
$1^-, 2^+, 3^+$. Alloantigen-activated CTL have been shown to be distinctly cross
reactive on TNP-modified syngeneic targets (42). With the split well technique,
this cross reactivity was shown to be hapten-specific (Fig. 2). H-2^b lymphnode
cells were activated with H-2^d spleen cells. At a concentration of responding cells
at which each well shows maximal activity on H-2^d target cells, some of the wells
are also cytotoxic for hapten-modified H-2^b cells. With two haptens, 2, 4, 6-tri-
nitrophenyl (TNP) and 3-nitro-4-hydroxy-phenyl (NP), only minimally cross reactive, one finds that clones activated by H-2^d can be toxic either to H-2^b -NP or
to H-2^b -TNP, but not for both (left panel). If activation is done with TNP-modified H-2^b cells, then one detects a few cross-reacting clones (right panel). Conversely, there is also evidence that CTL activated against virus-infected syngeneic

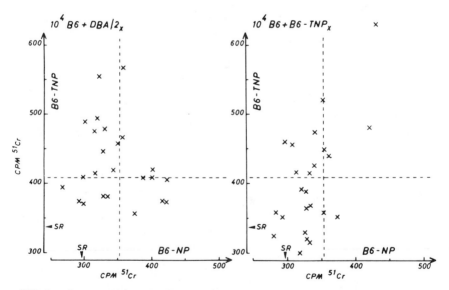

FIG. 2. Cross reactivity of cells from single microcultures. Comparison of the cytotoxic
activity of each half of the sensitized cells from single microcultures, divided on the day of
assay. Twenty-four cultures with 1×10^4 C57BL/6 lymph node cells were sensitized to allogeneic DBA/2 cells (left panel) or to TNP-modified syngeneic cells (right panel) and assayed
with NP-modified (abscissa) or TNP-modified (ordinate) C57BL/6 LPS-stimulated target
cells. Spontaneous release is indicated on the figure (SR); maximum release for the B6-NP
targets was 718 ± 68 cpm, for the B6-TNP targets 918 ± 45 cpm. The dashed horizontal
lines are three standard deviations above and to the right of the mean of the spontaneous
release values for the respective targets, the definition of positive cytotoxic activity.

cells (43) or H-2 identical target cells differing at minor histocompatibility loci (44) cross react with allogeneic cells. It thus appears that the populations of CTL reactive with allogeneic cells and with "self-H-2 + X" are at least overlapping, if not identical.

CTL RECEPTOR MODELS

According to the dual recognition hypothesis (45), H-2 restricted cytolysis requires two receptors, one with low affinity for self-H-2 and one for X, which can be a minor histocompatibility antigen, a virus or a hapten; allogeneic killing either proceeds via the antiself receptor, which might have a higher affinity for allogeneic H-2 (46), via the anti-X receptor, or via a third receptor, perhaps on a different cell altogether. Alternatively, the altered self model postulates that T cells have one receptor with a single antigen-binding site (possibly made up of more than one polypeptide chain), which recognizes the target antigen as a unit, whether a molecular complex of "self-H-2 + X," as here defined, or an allogeneic H-2. In this model, there is no recognition of self as such, but only of the complex, i.e., of altered self. Alloantigens represent self that has been altered genetically during evolution. The similarity of alloreactions and H-2 restricted cytolysis might extend even further, since it is conceivable that the CTL recognize not the allogeneic H-2 antigens as such, but rather as complexes formed of allogeneic H-2 molecules and any one of the non-H-2 surface antigens on allogeneic cells (47).

With the exception of chemically modified cells, where direct coupling of haptens onto H-2 molecules has been demonstrated (48), it is more likely that altered self antigens are generated by complexing or aggregation of cell surface molecules. Such association has been shown for Friend virus antigens and H-2D^b molecules (49). The association occurs exclusively with H-2D^b, not with H-2K^b, H-2D^d, or H-2K^d, and it is striking that T-cell-mediated cytotoxicity against Friend virus-infected cells is also restricted exclusively to H-2D^b (50). Similar restriction of cytolysis by one region only of a particular H-2 haplotype is known in the H-Y system (21) and could be explained by analogous limitations on complex formation.

The altered self hypothesis has the attraction of unifying the two manifestations of cytotoxicity into a single mechanism. Since there is no difference in principle between modified self and modified nonself, the model predicts the existence of cells specifically reactive in an H-2 restricted manner to haptenated allogeneic cells. Using the experimental model described in Fig. 3, we could show that after negative selection to remove those A cells that are responsive to B as such, the remaining A cells contained a subpopulation that could be stimulated by TNP-modified B to kill only TNP-B, not B alone, nor TNP-A (52).

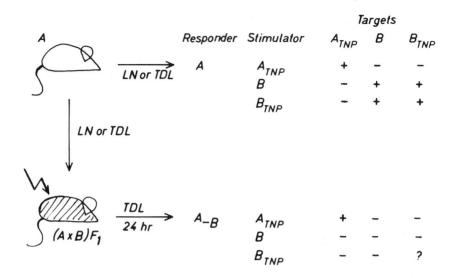

				Targets	
	Responder	Stimulator	A_{TNP}	B	B_{TNP}
	A	A_{TNP}	+	–	–
		B	–	+	+
		B_{TNP}	–	+	+
	A-$_B$	A_{TNP}	+	–	–
		B	–	–	–
		B_{TNP}	–	–	?

FIG. 3. Schematic presentation of experiment designed to detect cells specifically reactive to altered nonself. The strong reaction of A lymphocytes to B alloantigens normally prevents detection of a response specific for modified B (top half of figure). By recirculation through an irradiated $(A \times B)F_1$ host, the A lymphocytes reactive to B as such are specifically removed (51), and the remaining A cells can now be tested for their ability to respond to modified B after sensitization *in vitro*. The experiment was done both in rats and mice and showed clear responses of A-$_B$ to B-TNP (52).

Thus, under certain conditions, an H-2 restricted response to modified nonself is possible. Such a reaction is, however, critically dependent on the nature of the modification (52a). One explanation for this could be that CTL receptors may not have as much discriminatory power in the case of altered nonself as for altered self. Therefore, if the modification is too slight, potentially responding cells are lost in the negative selection by which it was intended to remove only specifically alloreactive cells.

If H-2 is viewed as a "fire alarm" and CTLs as "firemen," more likely for virus-infected than for transformed cells, then CTLs should be attuned to detect even minor modifications of self, and one would expect the repertoire to emphasize the latter. In the development of the receptor repertoire, the immune system must therefore steer between the Scylla of becoming autoreactive and the Charybdis of recognizing only alterations of self so extensive as to be unlikely. As predicted by Jerne (53), this adjustment of the repertoire depends on the antigenic environment, in particular the thymus, in which the T lymphocytes mature (54). Thus, A lymphocytes developing in an $A + B$ double bone marrow chimera can respond to $A + X$ or $B + X$, i.e., the A population as a whole can be restricted by two different H-2 types, only one of which is self (55-57). On the other hand, $(A \times B)F_1$

lymphocytes that develop in an A environment will respond only to $A + X$ and do not develop responsiveness to $B + X$, although B also represents self-H-2 (54, 58). Alloreactive cells either express an unchanged germline repertoire, or their reactivity is based on the cross reaction of "self + X" with alloantigens.

RUMINATIONS ON COMPONENTS OF CTL RECEPTORS

Which components are then likely to make up the T-cell receptor? As we have learned from B cells, an antigen-binding site constructed of two variable chains is an economical way to generate far more diversity than could be obtained with a single variable chain. It now seems reasonable to assume that in T cells at least one of these chains is encoded by immunoglobulin V genes.

Evidence is accumulating for the presence of immunoglobulin idiotypes on T helper and suppressor cells (59, 60) and on GVH and MLC-reactive T cells (61), as well as for H-chain-associated idiotypes and V_H allotypes on isolated receptors from otherwise unspecified, immune T cells (62, 63). Preliminary evidence for the presence of immunoglobulin idiotype on cytotoxic T cells has been referred to (61), but so far no formal proof exists. That T-cell-mediated lysis is unaffected by antiimmunoglobulins (8, 11) need not be significant, since such reagents do not necessarily react with those determinants of immunoglobulin that are relevant for T-cell receptors.

Assuming that the antigen-binding site of the T cell receptor is encoded by immunoglobulin V genes, one must then account for the facts that the great majority of T cells are alloreactive and that many examples of T-cell recognition of conventional antigens show H-2 restriction (reviewed in 64). Of course, specific alloantibodies can react with H-2, but there is no evidence that immunoglobulins in general are H-2 specific. T cells therefore either employ a very selected part of the V-gene repertoire [this could be the basis for the major idiotypes (64, 65)], or they employ a second kind of chain to make the receptor H-2 restricted.

Contrary to the findings of Binz et al. (66), Krammer and Eichmann (67) presented preliminary evidence that both the H-2 complex and the heavy chain linkage group determine expression of idiotype on MLC blasts. The role of H-2 in this case could be to influence the development and selection of the T-cell repertoire as discussed, rather than to code for a component of the receptor. However, Lonai et al. recently showed that antigen binding to Ly-1^+ T cells activated by a macrophage factor was inhibited by anti-V_H, antiidiotypic, and anti-Ia sera, while antigen binding by interferon-activated Ly-2^+, 3^+ T cells was inhibited only by anti-V_L and anti-H-2K and D sera (68, 69). These reports resurrect the idea that the MHC complex codes for part of the T-cell receptor (59, 70-72), and in particular they imply that the CTL receptor is made up of V_L and H-2. If that should be the case, then one requires an explanation for the conspicuous lack of effect on T-cell function in vitro of antisera against MHC products present on the responding T cells (73, 74), in particular for the lack of effect of anti-H-2 sera on CTL (Fig. 1; 32, 75, 76). Hopefully, we shall not have to wait long for the direct isolation and biochemical analysis of cell surface receptors from cytotoxic T lymphocytes.

CONCLUDING STATEMENT

CTL carry their own surface receptors that allow them to bind specifically to target cells of appropriate H-2 type. Binding occurs to the serologically defined H-2K (or D) molecules, both in allogeneic reactions and in H-2 restricted reactions to "self + X." The CTL populations reactive in these two cases are at least partially overlapping. An altered self model of CTL recognition would unify allogeneic and "self + X" cytolysis. The binding site of the receptor for altered self could in part be coded for by V gene(s), but could also contain components (not found in B cells) conferring H-2 specificity to the receptor.

Acknowledgments

I thank the European Molecular Biology Organization for a long-term fellow-ship. These studies were further supported by the Deutsche Forschungsgemein-schaft through Sonderforschungbereich 74.

References

1. Golstein, P., and Blomgren, H. *Cell. Immunol.* 9, 127, 1973.
2. Zagury, D., Bernard, J., Thierness, N., Feldman, M., and Berke, G. *Eur. J. Immunol.* 5, 818, 1975.
3. Katz, D. H. *In* "Lymphocyte Differentiation, Recognition, and Regulation." Academic Press, New York, 1977.
4. Perlmann, P., and Holm, G. *Adv. Immunol.* 11, 117, 1969.
5. Forman, J., and Möller, G. *J. Exp. Med.* 138, 672, 1973.
6. Bevan, M. J., Langman, R., and Cohn, M. *Eur. J. Immunol.* 6, 150, 1976.
7. Kuppers, R. C., and Henney, C. S. *J. Exp. Med.* 143, 684, 1976.
8. Brondz, B. D. *Transplant. Rev.* 10, 112, 1972.
9. Berke, G., Gabison, D., and Feldman, M. *Eur. J. Immunol.* 5, 813, 1975.
10. Nagy, Z., Elliott, B. E., and Nabholz, M. *J. Exp. Med.* 144, 1545, 1976.
11. Cerottini, J.-C., and Brunner, K. T. *Adv. Immunol.* 18, 67, 1974.
12. Elliott, B. E., Nagy, Z., Nabholz, M., and Pernis, B. *Eur. J. Immunol.* 7, 287, 1977.
13. Bach, F. H., Segall, M., Zier, K. S., Sondel, P. M., Alter, B. J., and Bach, M. L. *Science* 180, 403, 1973.

14. Peck, A. B., Alter, B. J., and Lindahl, K. F. *Transplant. Rev.* **29**, 189, 1976.
15. Kimura, A. K., Rubin, B., and Andersson, L. *Scand. J. Immunol.* **6**, 787, 1977.
16. Leclerc, J. C., Plater, C., and Fridman, W. H. *Eur. J. Immunol.* **7**, 543, 1977.
17. Alter, B. J., Schendel, D. J., Bach, M. L., Bach, F. H., Klein, J., and Stimpfling, J. H. *J. Exp. Med.* **137**, 1303, 1973.
18. Lindahl, K. F., and Bach, F. H. *Nature* **254**, 607, 1975.
19. Lindahl, K. F. Specificity of xenograft reactions *in vitro*. Ph.D. thesis, University of Madison, Wisconsin, 1975.
20. Bevan, M. J. *Nature* **256**, 419, 1975.
21. Simpson, E., and Gordon, R. D. *Transplant. Rev.* **35**, 59, 1977.
22. Doherty, P. C., and Zinkernagel, R. M. *J. Exp. Med.* **141**, 502, 1975.
23. Shearer, G. M. *Eur. J. Immunol.* **4**, 527, 1974.
24. Forman, J., and Vitetta, E. S. *Proc. Natl. Acad. Sci. USA* **72**, 3661, 1975.
25. Golstein, P., Kelly, F., Avner, P., and Gachelin, G. *Nature* **262**, 693, 1976.
26. Bevan, M. J., and Hyman, R. *Immunogenetics* **4**, 7, 1977.
27. Klein, J., Chiang, C. L., and Hauptfeld, V. *J. Exp. Med.* **145**, 450, 1977.
28. Billings, P., Burakoff, S., Dorf, M. E., and Benacerraf, B. *J. Exp. Med.* **145**, 1387, 1977.
29. Klein, J. "Biology of the Mouse Histocompatibility-2 Complex." Springer Verlag, New York, 1975.
30. Lindahl, K. F., Peck, A. B., and Bach, F. H. *Scand. J. Immunol.* **4**, 541, 1975.
31. Brunner, K. T., Mauel, J., Cerottini, J. C., and Chapuis, B. *Immunology* **14**, 181, 1968.
32. Nabholz, M., Vives, J., Young, H. M., Meo, T., Miggiano, V., Rijnbeek, A., and Shreffler, D. C. *Eur. J. Immunol.* **4**, 378, 1974.
33. Lemke, H., Hammerling, G. J., Hohmann, C., and Rajewsky, K. *Nature* **271**, 249, 1978.
34. Lindahl, K. F., Lemke, H., *Eur. J. Immunol.*, in press.
35. Lindahl, K. F., and Wilson, D. B. *J. Exp. Med.* **145**, 500, 1977.
36. Alter, B. J., Grillot-Courvalin, C., Bach, M. L., Zier, K. S., Sondel, P. M., and Bach, F. H. *J. Exp. Med.* **143**, 1005, 1976.
37. Lindahl, K. F., and Wilson, D. B., *J. Exp. Med.* **145**, 508, 1977.
38. Cantor, H., and Boyse, E. A. *Cold Spring Harbor Symp. Quant. Biol.* **41**, 23, 1976.
39. Skinner, M. A., and Marbrook, J. *J. Exp. Med.* **143**, 1562, 1976.
40. Teh, H. S., Harley, E., Phillips, R. A., and Miller, R. G. *J. Immunol.* **118**, 1049, 1977.
41. Hodes, R. J., Hathcock, K. S., and Shearer, G. M. *J. Immunol.* **115**, 1122, 1975.
42. Lemonnier, F., Burakoff, S. J., Germain, R. N., and Benacerraf, B. *Proc. Natl. Acad. Sci. USA* **74**, 1229, 1977.
43. Zweerink, H. J., Courtneidge, S. A., Skehel, J. J., Crumpton, M. J., and Askonas, B. A. *Nature* **267**, 354, 1977.
44. Bevan, M. J. *Proc. Natl. Acad. Sci. USA* **74**, 2094, 1977.
45. Doherty, P. C., Gotze, D., Trinchieri, G., and Zinkernagel, R. M. *Immunogenetics* **3**, 517, 1976.
46. Janeway, C. A., Binz, H., and Wigzell, H. *Scand. J. Immunol.* **5**, 993, 1976.
47. Matzinger, P. and Bevan, M. *J. Cell. Immunol.* **29**, 1, 1977.
48. Forman, J., Vitetta, E. S., Hart, D. A., and Klein, J. *J. Immunol.* **118**, 797, 1977.
49. Bubbers, J. E., and Lilly, F. *Nature* **266**, 458, 1977.
50. Blank, K. J., and Lilly, F. *Nature* **269**, 809, 1977.
51. Ford, W. L., and Atkins, R. C. *Nature (New Biol.)* **243**, 178, 1971.
52. Wilson, D. B., Lindahl, K. F., Wilson, D. H., and Sprent, J. *J. Exp. Med.* **146**, 361, 1977.
52a. von Boehmer, H., Haas, W., and Pohlit, H. *J. Exp. Med.*, 147, 1291, 1978.
53. Jerne, N. K. *Eur. J. Immunol.* **1**, 1, 1971.

54. Zinkernagel, R. M., Callahan, G. N., Althage, A., Cooper, S., Klein, P., and Klein, J. *J. Exp. Med.* **147**, 882, 1978.

55. von Boehmer, H., and Haas, W. *Nature* **261**, 141, 1976.

56. Zinkernagel, R. M. *J. Exp. Med.* **144**, 933, 1976.

57. Pfizenmaier, K., Starzinski-Powitz, A., Rodt, H., Röllinghoff, M., and Wagner, H. *J. Exp. Med.* **143**, 999, 1976.

58. Bevan, M. J. *Nature* **269**, 417, 1977.

59. Rajewsky, K., and Eichmann, K. *Contemp. Top. Immunobiol.* **7**, 69, 1977.

60. Cosenza, H., Julius, M. H., and Augustin, A. A. *Immunol. Rev.* **34**, 3, 1977.

61. Binz, H., and Wigzell, H. *Contemp. Top. Immunobiol.* **7**, 113, 1977.

62. Krawinkel, U., Cramer, M., Imanishi-Kari, T., Jack, R. S., Rajewsky, K., and Mäkelä, O. *Eur. J. Immunol.* **7**, 566, 1977.

63. Krawinkel, U., Cramer, M., Mage, R. G., Kelus, A. S., and Rajewsky, K. *J. Exp. Med.* **146**, 792, 1977.

64. Lindahl, K. F., and Rajewsky, K. *In* "Defense and Recognition" (E. S. Lennox, ed.), *Intern. Rev. Biochem.* **22**, 97, 1979.

65. Krawinkel, U., Cramer, M., Melchers, I., Imanishi-Kari, T., and Rajewsky, K. *J. Exp. Med.* **147**, 1341, 1978.

66. Binz, H., Wigzell, H., and Bazin, H. *Nature* **264**, 640, 1976.

67. Krammer, P., and Eichmann, K. *Nature* **270**, 733, 1977.

68. Lonai, P., and Steinman, L. *Proc. Natl. Acad. Sci. USA* **74**, 5662, 1977.

69. Lonai, P., Ben-Neriah, Y., Steinman, L., and Givol, D. *Eur. J. Immunol.* **8**, 827, 1978.

70. Ceppellini, R. *In* "Progress in Immunology" (B. Amos, ed.), p. 973. Academic Press, New York, 1971.

71. Crone, M., Koch, C., and Simonsen, M. *Transplant. Rev.* **10**, 36, 1972.

72. McDevitt, H. O. *Transplant. Proc.* **5**, 1799, 1973.

73. Shevach, E. M., Lee, L., and Ben-Sasson, S. Z. *In* "Immune Recognition" (A. S. Rosenthal, ed.), pp. 627-649. Academic Press, New York, 1975.

74. Meo, T., David, C. S., and Shreffler, D. C. *In* "The Role of Products of the Histocompatibility Gene Complex in Immune Responses" (D. H. Katz and B. Benacerraf, eds.), pp. 167-168. Academic Press, New York, 1976.

75. Schmitt-Verhulst, A. M., Sachs, D. H., and Shearer, G. M. *J. Exp. Med.* **143**, 211, 1976.

76. Burakoff, S. J., Germain, R. N., Dorf, M. E., and Benacerraf, B. *Proc. Natl. Acad. Sci. USA* **73**, 625, 1976.

GENERATION OF TNP-SPECIFIC H-2-RESTRICTED MURINE CYTOTOXIC CELLS AS A FUNCTION OF TNP-CELL SURFACE PRESENTATION

Carla B. Pettinelli, Anne-Marie Schmitt-Verhulst, and Gene M. Shearer

Major histocompatibility complex (MHC)-restricted T-cell-mediated cytotoxicity has been demonstrated for virus-infected (1) and chemically modified (2) murine cells as well as for cells expressing minor transplantation antigens (3, 4). In all cases, the restriction was shown to map to the K and/or D regions of the H-2 complex. To investigate the immunogenic parameters involved in the cytotoxic response, it was of interest to use the trinitrophenyl (TNP) hapten as a "modifying agent." Such studies have permitted the investigation of the specificity of cytotoxic effectors generated by stimulating cells presenting TNP on the cell surface in different ways. For example, the TNP group can be covalently coupled to the cell surface, but separated from it by a β-alanylalanylglycyl tripeptide (5), in contrast to the direct coupling of the cell surface proteins to ϵ-amino groups by trinitrobenzene sulfonate (TNBS) (1). Furthermore, murine lymphocyte surfaces have been modified with the amphipathic TNP compound TNP-stearyl-dextran (TSD), which interacts with cell surfaces by noncovalent interaction with the lipid bilayer of cell membranes (6). We have recently demonstrated that the addition of TNBS-conjugated soluble proteins such as bovine gammaglobulin (TNP-BGG) or bovine serum albumin (TNP-BSA) to cultures of mouse spleen cells leads to the generation of cytotoxic effector cells that are both H-2 restricted and TNP-specific (7). A number of parameters associated with this response have been investigated, and are compared in Table 1 with the same parameters previously studied for cytotoxic responses generated by sensitization with TNBS-modified syngeneic cells.

Murine spleen cells from normal donors were cultured *in vitro* with trinitro-

TABLE 1. Comparison of Various Parameters of the Cytotoxic Response Generated by Mouse Spleen Cells After Addition of TNP-Modified Syngeneic Cells of TNP Proteins

Parameters studied	Stimulation provided by addition of	
	TNBS-modified cells	TNP protein
Generation of primary *in vitro* cytotoxic response	Yes	Yes
TNP-dependence of response	Yes	Yes
H-2K and *H-2D* restricted specificity	Yes	Yes
Ir-gene-controlled preferential reactivity to *H-2K*^{k}	Yes	Yes
T-cell effected lysis	Yes	Yes
Generation of the response is dependent on glass-adherent cells	Yes	Yes

benzene sulfonate (TNBS)-conjugated soluble proteins, i.e., bovine γ-globulin (TNP-BGG) or bovine serum albumin (TNP-BSA). Addition of either of those two TNP-proteins to the spleen cell cultures led to the generation of cytotoxic T-effector cells that were *H-2* restricted and TNP specific. The lytic potential of such effectors was comparable to that generated by sensitization with TNBS-modified syngeneic cells, and it was restricted to haplotypes shared at the *K* or *K* plus *I-A*, or the *D* regions of the *H-2* complex. Greater effector cell activity was generated against TNBS-modified targets sharing *K* plus *I-A* than against modified targets sharing the *D* region with the responding cells. Such targets were modified by the addition of TNP-BGG to B10.BR spleen cells. This suggests that the same immune response gene is involved when the response is generated by the addition of either TNP-conjugated *soluble* proteins or TNBS-modified cells. The effector cells resulting from the addition of the TNP-conjugated soluble proteins were sensitive to rabbit antimouse brain serum and complement, which suggests that they are T lymphocytes. The generation of the cytotoxic effectors sensitized by the addition of TNBS-modified cells on TNP proteins is dependent on the presence of both T lymphocytes and a population of glass-adherent, radioresistant cells. The response did not appear to be attributable to antibody-dependent cellular cytotoxicity (ADCC), since TNP-lysine did not affect the lytic phase of the response under conditions that block ADCC.

    H-2-restricted, TNP-specific effector cells could be generated by culturing mouse spleen cells with syngeneic cells that had been preincubated with TNP_{40}-BGG or TNP_{30}-BSA for 1.5 hr. Addition of the unconjugated soluble proteins to the cultures did not result in cytotoxic effectors detectable on *H-2* matched targets, irrespective of whether the targets were prepared by modification with TNBS or by incubation with either the unconjugated or TNP-conjugated proteins. Target cells sensitive to lysis by effectors generated via sensitization with either TNBS cells or TNP protein could be prepared by preincubation with TNP proteins for as little as 10 min. Depletion of phagocytic cells in the tumor preparation by Sephadex G-10 column fractionation prior to incubation with TNP-BSA had no effect on their lysis as targets by the relevant effector cells. There was no

TABLE 2.   Lack of Evidence for "Carrier" Protein Specificity of *H-2*-Restricted Cytotoxic Effector Cells Generated by the Addition of TNP Proteins

Responding cells stimulated by addition of	Target cells preincubated with	Lysis detected
TNP-BGG	TNP-BGG	Yes
TNP-BSA	TNP-BSA	Yes
TNP-BGG	TNP-BSA	Yes

evidence for the involvement of the "carrier" protein in the specificity of the effector cells generated by TNP-BSA or TNP-BGG, since the effectors generated by addition of TNP-BSA or TNP-BGG lysed syngeneic tumor target cells preincubated with TNP-BSA (Table 2).

At least three mechanisms have been considered that could account for the generation of *H-2*-restricted, TNP-specific, cytotoxic T-effector cells by the addition of soluble TNP proteins. These include: (a) covalent linkage of activated TNP groups from the soluble proteins to cell surface components; (b) macrophage processing of the soluble conjugates and their presentation to the responding lymphocytes in association with *H-2*-coded self-structures; (c) hydrophobic interaction of the TNP proteins to cell surfaces. However, results obtained from SDS gel patterns (7) that indicate that spleen cell-bound TNP was still linked to BSA, and the observation that phagocytic-depleted cells could interact with the soluble TNP proteins and function as *H-2*-restricted targets appear not to favor the proposed mechanisms (a) and (b). However, phagocytic cells are required for the stimulation of cytotoxic effectors in the presence of either TNBS-modified cells or TNP proteins.

The analysis of the antigenicity and the immunogenicity of TNP-presenting cells as a target or stimulator for TNP-specific, *H-2*-restricted T-cell mediated cytotoxicity is summarized in Table 3 as a function of the mode of presentation of the TNP moiety on the cell surface. Direct TNBS reaction of ϵ-amino groups of

TABLE 3.   Comparison of Cytotoxic Responses Detected as a Function of TNP Presentation on Stimulating and Target Cells

Method of TNP presentation on			
Stimulating cells	Target cells	Lysis of target cells	Reference cited
TNBS	TNBS	Yes	2
TNBS	TNP-AGG	No	5
TNBS	TSD	No	6
TNBS	TNP protein	Yes	7
TNP-AGG	TNP-AGG	Yes	5
TNP-AGG	TNBS	Marginal	5
TSD	TSD	No	6
TSD	TNBS	No	6
TNP protein	TNP protein	Yes	7
TNP protein	TNBS	Yes	7

cell surface proteins generates efficient stimulating cells for syngeneic responders. The effector cells resulting from such a sensitization lyse both TNBS-modified syngeneic target cells and target cells preincubated with TNP proteins; however, target cells on which the TNP moiety is bound to the β-alanyl group of the tripeptide spacer alanylglycylglycyl, which is covalently bound to cell surface proteins, are not lysed (5). Furthermore, target cells presenting TNP coupled to stearyl-dextran noncovalently inserted in the cell membrane are not lysed (6). Effector cells generated when cultures were sensitized with TNP-AGG-modified syngeneic cells lysed H-2-matched, TNP-AGG-modified targets but only marginally lysed H-2 matched TNBS-modified targets (5). TSD-modified cells were not effective either in generating TNP-specific cytotoxic effector cells or in serving as targets for effectors generated by sensitization with TNBS-modified cells (6). In contrast, cells preincubated with TNP-BSA or TNP-BGG were effective as stimulating or target cells for TNP-specific, H-2-restricted cytotoxicity.

These results indicate that simultaneously presentation of the relevant syngeneic H-2-coded product and of the TNP moiety on the target cell surface is not sufficient to generate a TNP-specific antigenic structure capable of being recognized by effectors sensitized to TNBS-modified cells. The lysis of target cells presenting the TNP moiety couple to the carrier proteins BSA or BGG on their surface (7) indicates, however, that a covalent modification of the cell surface proteins (and thus H-2-coded products) by TNP is not an absolute requirement for the recognition by effectors sensitized by TNBS-modified cells. The failure of TSD to present the TNP moiety on the cell surface in an immunogenic fashion to T-cytotoxic precursor cells could be accounted for in at least two ways: (a) TSD may place the TNP groups in an unfavorable position with respect to the relevant self-components required for sensitization (e.g., H-2-coded products); or (b) stearyl dextran may not be an immunogenic carrier for sensitization, whereas cell surface through the alanylglycylglycyl spacer is immunogenic, but the effector cells that are generated are not able to lyse TNBS-modified syngeneic target cells. At least two interpretations can be given to account for the lack of lysis of TNP-AGG target cells by effectors sensitized by TNBS-modified cells in contrast to the lysis of TNP-BSA preincubated targets by the same effectors. First, the specificity associated with recognition of the TNP group could be dependent on the number of methylene carbons in the adjacent amino acid. The predominant group reacting with TNBS on the cells and BSA would be the ϵ-amino group of lysine. In contrast, in the TNP-AGG structure, TNP is linked to the β-amino group of alanine. This might also account for the lack of lysis of TSD-modified targets by effectors generated via sensitization with TNBS-modified cells. Secondly, it could be that TNBS-modified cell surface proteins and TNP-conjugated BSA or BGG associated with the cell surface interact with the relevant H-2-coded product in a similar fashion, whereas the presentation of TNP through the AGG spacer relative to the H-2-coded product is fixed in a different conformation.

CONCLUDING STATEMENT

The observations summarized here broaden the limits of the experiments and interpretations imposed by covalent modification with TNBS to include TNP groups attached to soluble protein carriers. Thus, the chemically modified, *H-2*-restricted CML model may not depend on actual covalent conjugation of the TNP group to *H-2*-coded cell surface products, and therefore could be similar to the *H-2*-restricted viral and minor transplantation antigen models. These findings could also raise the possibility that *H-2*-restricted cytotoxic immune reactions involving autoimmunity may be generated by soluble components associated with autologous cells.

References

1. Doherty, P. C., Blanden, R. V., and Zinkernagel, R. M. *Transplan. Rev.* **29**, 89, 1976.
2. Shearer, G. M., Rehn, T. G., and Schmitt-Verhulst, A.-M. *Transplan. Rev.* **29**, 222, 1976.
3. Rehn, T. G., Inman, J. K. and Shearer, G. M. *J. Exp. Med.* **144**, 1134, 1976.
4. Gordon, R. D., Simpson, E., and Samelson, L. E. *J. Exp. Med.* **142**, 1108, 1975.
5. Bevan, M. J. *Nature* **256**, 419, 1975.
6. Henkart, P. A., Schmitt-Verhulst, A.-M., and Shearer, G. M. *J. Exp. Med.* **146**, 1068, 1977.
7. Schmitt-Verhulst, A.-M., Pettinelli, C. B., Henkart, P. A., Lunney, J. K., and Shearer, G. M. *J. Exp. Med.* **147**, 352, 1978.

SECONDARY SENDAI-SPECIFIC T-EFFECTOR CELLS: REQUIREMENTS FOR RESTIMULATION, TARGET CELLS RECOGNITION, AND LYSIS

Ulrich Koszinowski, Mary Jane Gething, Evan T. Smith, and Michael Waterfield

Virus-infected cells may become targets of immunological attack by T cells after insertion of virus proteins into the host plasma membrane (1). Similarly, modification of cells with haptens and sensitization to minor H antigens result in generation of specific T-effector cells (for review, see 2). The T-effector cells raised by these three types of antigenically altered cells have one property in common: Lysis requires not only expression of the sensitizing antigen on the target cell plasma membrane but also homology between attacker and target cells restricted by the MHC.

An investigation of the viral protein expressed on the host cell membrane is undertaken to determine the specificity of antigenic determinants recognized by the T-effector cells and their precursors.

This report focuses on three questions:

(a) How are viral antigens presented to form target cells recognized by T cells?

(b) Does virus-specific adsorption of T-effector cells to infected cell monolayers have the same antigenic requirements as cytolysis?

(c) Are macrophages capable of uptake and presentation of viral antigens associated with host cell proteins?

RESULTS AND DISCUSSION

Sendai virus was used throughout this study for several reasons: (a) the mouse is a natural host for Sendai virus; (b) after infection, cell-mediated immunity can be demonstrated *in vitro* (3,4); (c) Sendai virus expresses two glycoproteins on host cell surfaces as well as on the virus envelope, the haemagglutinin-neuraminidase (HANA), and the fusion factor (F), which is associated with virus-cell and cell-cell fusion; (d) Sendai virus retains its fusion capacity after inactivation of infectivity.

In an earlier publication (5), we described how Sendai virus inactivated with UV or β-propiolactone (6) was still capable of expressing antigens on the cell surface recognized by T cells. In addition, artificial liposomes could be used for insertion of the two glycoproteins into the target cell membrane.

It has been suggested that MHC antigens may be altered during the process of cell infection (1). Our results exclude any alteration of host or viral proteins by "modification from within," i.e., modification at the level of transcription, translation, or posttranslational glycosylation. However, the possibility remains that there may be "modification from without," i.e., enzyme or steric alteration of cell membrane proteins by viral activity at the cell surface.

An additional series of experiments indicates that attachment of virus to the cell membrane is not sufficient for target cell formation and that fusion of the viral glycoproteins is required for Sendai-specific T-cell lysis. Thus, Sendai virus grown in MDBK cells was ineffective in target cell formation. This virus contains a precursor fusion factor (F_0) and is therefore inactive in fusion and noninfectious, although it contains an active haemagglutinin-neuraminidase protein and can attach to the cell surface (7,8). Activation of the fusion factor requires proteolytic cleavage of the precursor F_0. When Sendai virus is grown in embryonated eggs, this cleavage occurs at the plasma membrane. In MDBK cells the cleavage does not occur but virus grown in MDBK cells can be cleaved *in vitro* by low concentrations of trypsin with restoration of fusion and haemolytic activity (9). This cleaved virus was highly effective in target cell formation. These results are shown in Table 1. The residual target cell formation with MDBK-Sendai (about 0.1% of the activity of cleaved virus) is due to a small proportion of cleaved F that was always found in the virus preparations.

A further approach to this question was made using a variety of proteolytic enzymes such as trypsin, chymotrypsin, pronase, and the staphylococcus aureus v8 protease to inactivate either one or both of the glycoproteins. In confirmation of these results, we have shown that attachment of virus to cells is by itself not sufficient for target cell formation. Thus, when virus is treated with trypsin (at higher concentrations than those required for F_0 cleavage) the fusion activity is lost, whereas haemagglutinin and neuraminidase activity is retained. Attachment of such virus to cells can be demonstrated using a radioimmune assay, but this does not lead to T-cell-mediated target cell lysis. Virus treatment with chymo-

TABLE 1. Cleavage of the Precursor Protein (F_0) to Active Fusion Factor Confers to Sendai Virus in Ability to Form Target Cells

Virus incubated with 10^6 P-815 cells[a]		Specific ^{51}Cr release Ratio K:T		Haemolysis assay of virus preparations[b]	E_{549}
		20:1	5:1		
Egg-grown Sendai virus	1 μg	92.5	89.3	Egg grown Sendai virus 20 μg	1.44
MDBK-Sendai virus	100 μg	16.0	14.7	MDBK-Sendai virus	
MDBK-Sendai virus	10μg	5.3	4.6	20 μg	0.07
Egg-grown Sendai virus trypsin treated	1 μg	82.8	76.3		
MDBK-Sendai virus trypsin treated	100 μg	79.3	76.5	MDBK-Sendai virus	1.19
MDBK-Sendai virus trypsin treated	10 μg	7.0	2.8	20 μg	
MDBK-Sendai virus trypsin treatment control	100 μg	15.8	11.0		

[a] Preparation of target cells and cytotoxic test as described (5).
[b] Haemolytic activity assayed by the method of Shimizu et al. (9).

trypsin and *S. aureus* protease results in inactivation of the HANA glyprotein, and pronase treatment cleaves both glycoproteins; after both treatments, target cell formation was significantly reduced. Since virus cell fusion mediated by the active fusion factor requires prior attachment via the haemagglutinin, it is not possible to determine which of the F and HANA glycoproteins contributes the virus specificity of T-cell recognition.

The adsorption characteristics of secondary anti-Sendai cytolytic cells were determined using syngeneic and allogeneic fibroblast monolayers, that were either noninfected or infected with Sendai virus (Table 2, expts. 1-3). Two distinct levels of adsorption were observed. When presented with noninfected adsorbents, syngeneic or allogeneic, anti-Sendai cells exhibited a considerable degree of non-specific adsorption (NAS), a common feature of fibroblast adsorption systems (10; Smith and Golstein, in preparation). Significantly higher levels of adsorption were obtained, however, if the monolayers were infected with Sendai. Such adsorption was not H-2 restricted but occurred to the same extent on syngeneic and allogeneic fibroblasts. This may be interpreted as evidence for antigen-specific (AS) adsorption in which case it contrasts with the H-2 restriction of cytolytic activity. Alternatively, it may not be related to the recognitive events of cytolysis but simply reflect some nonspecific adherence mechanism induced by virus infection that enhances NAS adsorption to the infected monolayers. In an attempt to distinguish between these possibilities, three further types of experiments were carried out.

Experiment 4 of Table 2 addressed the temperature dependence of adsorption. In mouse antiallogeneic cytolytic systems, AS adsorption is strongly temperature-dependent occurring at 37°C but not at 4°C (reviewed in 11), whereas NAS

TABLE 2. Adsorption of Anti-Sendai Cytolytic Cells to Fibroblast Monolayers

Experiment No.	Nature of cytolytic cells[a]	Unfractionated control	Nonadsorbed fractions from					
			BALB/c monolayers[b]			C57B1/6 monolayers[b]		
			o	s	f	o	s	f
1	H-2^d anti-Sendai	51^c	41^c	26	—	37^c	28	—
			39	27	—	35	28	—
2	H-2^d anti-Sendai	64	11	2	—	13	1	—
			—	1	—	—	—	—
3	H-2^d anti-Sendai	49	19	3	—	24	4	—
			21	3	—	24	4	—
4	H-2^d anti-Sendai							
	A:40°C	23	16	17	—	18	16	—
	B:37°C	20	10	2	—	9	-1	—
5	H-2^b anti-H-2^d	26	8	8	—	18	18	—
			10	9	—	21	17	—
6	H-2^d anti-Sendai	58	34	24	37	34	28	35
	H-2^d antiinfluenza A^d	54	38	40	31	38	39	40

Equal volumes of the cytolytic cell suspension were incubated for 2 hr at 37°C (40°C in expt. 4A) on the appropriate fibroblast monolayers or in the absence of such monolayers (unfractionated control). The nonadsorbed cells were then harvested, resuspended in a standard volume and tested for cytotoxicity using appropriate targets. Adsorption is thus defined by the depletion of cytolytic activity in the nonadsorbed fractions. Comparable numbers of nonadsorbed cells were recovered from each type of monolayer (usually about 50% of the unfractionated control). The procedure was derived from Golstein et al. (10).

[a] H-2^d anti-Sendai, antiinfluenza: Secondary *in vitro* activated BALB/c spleen cells directed against syngeneic cells infected with Sendai and influenza virus.

[b] o: uninfected; s: fused with β-propylactone inactivated Sendai virus (40 μg/10 cm^2 monolayer); f: infected with influenza virus grown in chicken eggs (Jap/Bel, 10^6 pfu/10 cm^2 monolayer).

[c] % ^{51}Cr-release minus spontaneous release.

[d] Preliminary experiments are inconclusive with respect to the H-2 restriction of adsorption in the influenza system, in some experiments such restriction was not observed.

adsorption is usually enhanced at low temperature (Smith and Golstein, in preparation). It was found that the adsorption of anti-Sendai cells did exhibit such a pattern of temperature dependence. However, it might still be argued that a non-specific virus-dependent adherence mechanism could also be temperature dependent, e.g., cell fusion mediated by Sendai virus.

In the second type of experiment (expt. 5 of Table 2), appropriate anti-allogeneic cells were applied to the four types of monolayer in order to determine whether the Sendai virus-fused monolayers exhibited an enhanced capacity for NAS adsorption. No evidence for such a phenomenon was found: neither NAS (to C57Bl/6) nor AS (to BALB/c) adsorption was affected by virus infection of the monolayers.

In the final type of experiment, virus specificity was determined by including monolayers infected with influenza virus. Experiment 6 of Table 2 indicates that anti-Sendai cells exhibit only NAS adsorption to influenza-infected monolayers even though such syngeneic monolayers are capable of adsorbing analogous antiinfluenza cytolytic cells in an apparently AS manner. Hence, the higher level of adsorption of anti-Sendai cells to infected monolayers is specific for Sendai infection.

These results suggest that AS adsorption representing the initial recognitive events in the cytolytic process is responsible for the higher level of adsorption of anti-Sendai cells to Sendai-infected monolayers. If this first step is not H-2 restricted, unlike full cytolysis, a second recognitive event would be required for the completion of cytolysis. This would imply a two-step dual recognition model.

Infectious Sendai virus, inactivated virus, and even artificial liposomes containing the two glycoproteins, all boost T-cell restimulation when added to primed spleen cell cultures. It seems likely that virus in suspension will not boost the responder cell, but becomes attached and fused to cells in the culture that then display viral antigens together with host cell determinants. Unexpectedly, allogeneic cells, when fused with Sendai virus, are capable of boosting secondary T-effector cells for an H-2 restricted activity as well.

One possible explanation for this finding is that after breakdown of these stimulator cells the antigens are taken up and presented by responder type, i.e., syngeneic macrophages. We established a system to demonstrate these macrophage effects. Macrophages were eliminated from responder cells by passage through Sephadex G10 columns resulting in over 99% depletion of macrophages (13). To avoid nonspecific effects (13, 14) all cells were cultured in the presence of 2-mercaptoethanol. Under these conditions living stimulator cells fused with virus were still active in boosting a secondary T-cell response. Subsequently, the stimulator cells were inactivated either by heat treatment (45°C, 60 min) or by repeated freeze-thawing. These cells did not, or only marginally, boost macrophage-depleted responder cell cultures (Fig. 1). Addition of syngeneic normal macrophages in a ratio of 1-5×10^5 macrophages to 5×10^6 responder cells restored the capacity of these cultures to generate cytolytic T cells (Fig. 2). More than 5×10^5 macrophages per culture had inhibitory effects.

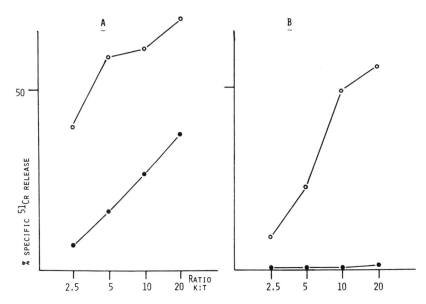

FIG. 1. Effects of stimulator cell inactivation and responder type macrophage depletion on secondary killer cell generation. ○, Responder cultures containing macrophages; ●, Responder cultures macrophage depleted. (A) BALB/c Con-A stimulated blast cells, fused with 20 µg inactivated Sendai virus /10^6 cells. (B) Inactivation of stimulator cells at 45°C for 1 hr. Harvest and test of T-effector cells after six days in culture; target: P-815 fused with 10 µg /10^6 cells Sendai virus.

Thus a system was established that allows the investigation of H-2 requirements for macrophage help in this system. Using different combinations of responder cells, stimulator cells, and macrophages, the following results were obtained (Fig. 3):

(a) Responder cells containing macrophages can be stimulated with living or heat-treated syngeneic as well as allogeneic cells provided Sendai virus antigens are present.

(b) Responder cells depleted of macrophages respond well to living syngeneic but poorly to inactivated syngeneic stimulators cells. Syngeneic as well as allogeneic macrophages significantly increase the generation of killer cells in these cultures.

(c) Responder cells depleted of macrophages respond to living but not to inactivated allogeneic stimulator cells.

Syngeneic macrophages added to the cultures with inactivated allogeneic stimulator cells will restore the T-cell response. But allogeneic macrophages that help to restore the response to inactivated syngeneic stimulator cells fail to help in the response to allogeneic stimulator cells carrying virus.

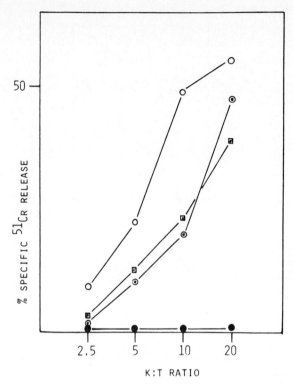

FIG. 2.   Restoration of secondary killer cell response by addition of macrophages. Respon-
der cells: 5×10^6 cells/well macrophage depleted BALB/c spleen cells primed *in vitro* to
Sendai one month earlier. ○, Viable stimulator cells: BALB/c Con-A blasts fused with virus;
stimulator responder ratio 1:10; ●, same stimulator cells after incubation 45°C, 1 hr;
⊙, 1×10^5 BALB/c macrophages added to cultures with inactivated stimulator cells; ⊡,
5×10^5 CBA macrophages added to cultures with inactivated stimulator cells.

There is no obvious explanation for the boosting effect of living allogeneic
virus-fused cells in the absence of macrophages. Perhaps the strong allogeneic
stimulus (15) activates memory cells in some nonspecific way. Another possibility is that there is antigenic stimulus by the viral antigen alone, provided it is
present in high concentration and integrated into a membrane of a living cell.
Under conditions where macrophages are required, there is a restriction by the
MHC. One explanation is that viral antigens can be presented by macrophages
independent of host cell H-2 or in association with host proteins. Thus allogeneic
macrophages can present antigens from inactivated syngeneic cells by copresen-
tation of a complex of host cell H-2 and viral antigen, but fail to do so when the
host H-2 is allogeneic as well. Syngeneic macrophages present the complex of
syngeneic host H-2 together with viral antigens but in the case of allogeneic
stimulator cells they seem to be able to present the viral antigens, together with
their own H-2, whether they are complexed with allogeneic H-2 or not.

It is too early to draw general conclusions about antigen presentation by

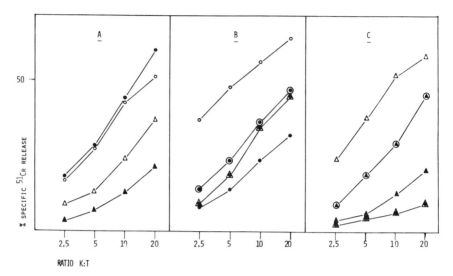

FIG. 3. H-2 restricted help of macrophages in the secondary response to different stimulator cells. (A) BALB/c responder cells not depleted of macrophages. Stimulator cells:
○, BALB/c-Sendai; ●, BALB/c-Sendai, heat inactivated; △, CBA-Sendai; ▲, CBA-Sendai, heat
inactivated. (B) BALB/c responder cells depleted of macrophages; syngeneic stimulator cells;
added macrophages: ○, 1 × 10⁵ CBA macrophages added to cultures with inactivated stimulator cells. △, 1 × 10⁵ CBA macrophages added to cultures with inactivated stimulator cells.
(C) BALB/c responder cells depleted of macrophages; allogeneic CBA stimulator cells; added
macrophages as in (B).

macrophages in H-2 restricted T-cell responses. In a series of experiments we tried
to investigate the presentation of H-Y antigens during secondary stimulation *in
vitro*. There was no T-cell activation in any of these combinations when inactivated
stimulator cells were used. Perhaps there is a difference in the capacity of macrophages to help in the response to those antigens.

Further experiments are required to map within the MHC the homology
required between responder cells, inactivated stimulator cells, and macrophages
and to investigate the mechanisms of macrophage help in this system.

CONCLUSIONS

Sendai virus provides a useful model for studying plasma membrane alterations due to insertion of foreign proteins. Since the experiments can be carried
out with inactivated virus or envelope proteins from inactivated virus, effects
due to insertion of proteins into plasma membranes can be separated from viral
effects on host cells due to infection. The use of purified viral glycoproteins in
the absence of viral infection allows a quantitative approach to plasma membrane

alterations. In addition, unlike the binding of haptens to lysine residues of cell surface proteins, there is no obvious direct alteration of host cell proteins after insertion of the Sendai virus glycoproteins. For target cell formation we could demonstrate that fusion of the glycoproteins into the cell membrane is required. Since both HANA and F glycoproteins are necessary for protein insertion into the plasma membrane (16), the definition of the protein introducing the viral specificity recognized by T cells has not yet been possible.

Experiments with secondary effector cells specific for Sendai virus adsorbed to infected cell monolayers *in vitro* gave unexpected results, differing from results obtained in similar experiments with effector cells specific for vaccinia virus (17) or influenza A virus. Adsorption was found to be virus-specific although not H-2 restricted. There may be a first phase recognition of viral antigens, different from recognition in other virus infections, due to some properties of Sendai virus, perhaps the fusion capacity.

The fusion factor activity also provides a possible explanation for the results of the macrophage dependent secondary stimulation of Sendai-specific effector cells *in vitro*. Results obtained so far suggest that both H-2 and viral antigen(s) are required for restimulation. It is our hypothesis that for effective restimulation allogeneic macrophages must present both host H-2 proteins and viral antigens. Since the same approach failed to induce secondary effector cells to minor H antigens, it is possible that the fusion factor enables macrophages to present the viral antigens or the virus-host cell protein complex in an antigenic form.

Acknowledgments

U. K. was supported by a grant of the Deutsche Forschungsgemeinschaft.

References

1. Zinkernagel, R. M., and Doherty, P. C. *Nature* **248**, 701, 1974.
2. Munro, A., and Bright, S. *Nature* **264**, 145, 1976.
3. Ertl, H., and Koszinowski, K. *Immunobiology* **152**, 128, 1976.
4. Schrader, J. W., and Edelman, G. M. *J. Exp. Med.* **145**, 523, 1977.
5. Koszinowski, U., Gething, M. Y., and Waterfield, M. *Nature* **267**, 160, 1977.
6. Neff, J., and Enders, J. F. *Proc. Soc. Exp. Bio.. Med.* **127**, 260, 1968.
7. Homma, M., and Ohuchi, M. *J. Virol.* **12**, 1457, 1973.
8. Scheid, A., and Choppin, P. W. *Virology* **57**, 475, 1974.
9. Shimizu, K., Hosaka, Y., and Shimizu, Y. K. *J. Virol.* **9**, 842, 1972.
10. Golstein, P., Svedmyr, E. A. J., and Wigzell, H. *J. Exp. Med.* **134**, 1385, 1971.
11. Golstein, P., and Smith, E. T. *Contemp Top. Immunol.* **7**, 1977.

12. Golstein, P. *Nature* **252**, 81, 1974.
13. Ly, I. A., and Mishell, R. I. *J. Immunol. Methods.* **5**, 239, 1974.
14. Chen, Ch., and Hirsch, J. G. *J. Exp. Med.* **136**, 604, 1972.
15. Schendel, D., Alter, B., and Bach, F. *Transplant Proc.* **5**, 1651, 1973.
16. Hosaka, Y. and Shimizu, K. *In* "Virus Infection and the Cell Surface" (Poste, G. and Nicolson, G. E., eds.), pp. 129-155. North-Holland Amsterdam, 1977.
17. Koszinowski, U., Ertl, H. Wekerle, H., and Thomssen, R. *Cold Spring Harbor Symp. Quant. Biol.* **XLI**, 529, 1976.

CYTOTOXIC T LYMPHOCYTES INDUCED BY H-2 NEGATIVE STIMULATOR CELLS

Hermann Wagner, Martin Röllinghoff, and Pierre Golstein

Most, if not all, immune functions of murine T cells are presently considered to be associated with H-2 gene products. Whereas the peripheral regions of the MHC (H-2K, H-2D) appear to be instrumental for cytotoxic T lymphocytes specific to foreign antigens, the I-region is critical for T helper cells, T cells proliferating in the MLC, and T cells transferring DTH reactions (1-7). In the case of CTL it has been argued that their lytic function can take place only on H-2-positive target cells. This proposition is based on results demonstrating that both virus specific and hapten specific H-2-restricted CTL are unable to lyse virus-infected or hapten-conjugated H-2-negative targets (8, 9). If, however, the specificity of CTL generated is determined by the antigen used for the induction of the immune response, antigen-specific CTL induced against H-2-positive, hapten-conjugated (or virus-infected) stimulator cells ought not to lyse H-2-negative, hapten-conjugated (or virus-infected) targets. If on the other hand, T cells must recognize H-2 gene products in addition to foreign antigens to become sensitized or to mediate lytic effector functions, it should not be possible to induce CTL with specificity for H-2-negative target cells. In the experiments to be described we explored this issue using an H-2-negative embryo carcinoma cell line (10) as stimulator cells.*

RESULTS AND DISCUSSION

It is known (8, 9) that H-2-restricted hapten or virus-specific murine CTL are unable to lyse *in vitro* hapten-conjugated or virus-infected target cells devoid

---

Editors note: The role of xenogeneic serum determinants in the induction of cytolysis against H-2-less embryonal F9 carcinoma cells is discussed by the authors in a publication that appeared in *J. Immunol*. 121, 2533, 1978.

TABLE 1. Lytic Activity of Anti-Sendai Virus, Anti-TNP and Anti-F9
Immune Effector Cells

	% Specific lysis of targets[a]					
Effector cells	F9	F9-TNP	F9-Sendai	$H\text{-}2^{bc}$ blasts	$H\text{-}2^{bc}$ blasts-TNP	$H\text{-}2^{bc}$ macrophages Sendai
$H\text{-}2^{bc}$ anti-129-TNP[b]	3	5	ND	4	37	ND
$H\text{-}2^{bc}$ anti-Sendai[c]	−1	ND	6	2	ND	29
$H\text{-}2^{k}$ anti-F9[d]	34	37	29	1	ND	ND
$H\text{-}2^{k}$ anti-$H\text{-}2^{bc}$[e]	2	0	3	67	ND	ND

[a] Ratio of effector to target cells of 50 : 1; 6 hr cytotoxicity assay; background lysis of
the target cells was less than 34%; blasts refer to LPS-induced (13) splenic blast lymphocytes. ND = not done.
[b] TNP-specific CTL were induced *in vitro* according the Shearer (1) as described (11)
using mouse 129 ($H\text{-}2^{bc}$) responder cells.
[c] Sendai virus-specific CTL were induced *in vivo* as described (12) using strain 129 mice
($H\text{-}2^{bc}$) cells.
[d] 4×10^6 CBA ($H\text{-}2^k$) splenic responder cells were cocultivated with 0.2×10^6 x-irradiated (2000 R) F9 cells as described (13) over a culture period of 5 days. Thereafter
the cells were assayed for cytotoxicity.
[e] $H\text{-}2^k$ anti-$H\text{-}2^{bc}$ CTL were induced in a MLC using CBA responder cells and mouse
129 derived stimulator cells under conditions previously described (13).

of H-2 cell surface antigens. In confirming this observation (Table 1), it became
apparent that these results could also be explained in terms of the known antigen specificity of the H-2 restricted CTL. Thus CTL triggered by H-2-positive,
hapten-conjugated (or virus-infected) stimulator cells might require recognition
of both the syngeneic H-2 antigens and the hapten or virus for specific target
cell lysis to occur. Consequently, to test whether CTL are capable to lyse H-2-
negative target cells, we sought to raise CTL to H-2-negative stimulator cells.
Typical results obtained are incorporated in Table 1 and clearly suggest that
CBA ($H\text{-}2^k$) splenic lymphocytes cocultured with x-irradiated F9 cells do, in
fact, generate cytotoxic lymphocytes capable of lysing ^{51}Cr-labeled F9 target
cells. It should be noted that the F9 target cells lysed by anti-F9 immune cytotoxic effector cells were not affected by anti-$H\text{-}2^{bc}$ reactive CTL, confirming
that the F9 cells used were $H\text{-}2^{bc}$ negative. The data given in Table 2 indicate
that the anti-F9 immune cytotoxic effector cells were T cells. Thus, responder
cells enriched for T lymphocytes were capable of generating cytotoxic effector
cells *in vitro*, and treatment of the effector cells with anti-Thy-1.2 serum plus
complement abrogated their cytotoxic activity (Table 2). From these data it
may be inferred that CTL precursors can be sensitized by H-2-negative stimulator cells resulting in CTL capable of recognizing and lysing H-2-negative target
cells. Operationally, H-2-restricted CTL (1-4, 11) and anti-F9 immune CTL
appear to differ with respect to the H-2 requirement for both the induction and
expression of their cytolytic effector functions; the effector activity of anti-F9

TABLE 2. Effect of Anti-Thy-1.2 Serum Plus Complement Upon
Anti-F9 Immune Cytotoxic Effector Cells

Treatment of effector cells[a]	% Specific lysis of F9 target cells	
	100 : 1	10 : 1
No treatment	42	27
Complement	42	26
Anti-Thy-1.2 serum	44	29
Anti-Thy-1.2 serum plus Complement	17	4

[a] Splenic lymphocytes of CBA mice injected sc 4 weeks previously with 2×10^6 x-irradiated (2000 R) F9 cells were treated with rabbit anti-mouse Ig antiserum plus complement in the presence of Na-azide in order to kill Ig-positive lymphocytes. After dead cell removal, the viable cells (4×10^6 responder cells) were cultured together with 0.25×10^6 x-irradiated F9 stimulator cells over 5 days. Thereafter various groups of cultured cells were treated either with complement, anti-Thy-1.2 serum or anti-Thy 1.2 serum plus complement and assayed for cytotoxicity against ^{51}Cr-labeled F9 cells.

CTL being H-2 independent in contrast to the functional activity of H-2K/D (1-4) or H-2I (14, 15) region restricted CTL.

What then could be the nature of the target antigen recognized on F9 cells? To this end we tested the cytotoxic activity of anti-F9 immune CTL against a variety of different target cells. From the data developed (Tables 3, 4) it can be seen that only targets carrying the F9 antigen were effectively lysed. Interestingly, CBA ($H-2^k$) anti-F9 CTL were capable of lysing syngeneic spermatogonia (Table 4), known to express the F9 antigen, and to be H-2 negative (10). Since mouse 129 anti-F9 antiserum, containing antibodies specific for F9 antigens (16), blocked the anti-F9 CTL (13), the data would suggest that anti-F9 CTL recognize the F9 antigen as the target antigen. However, it is conceivable that the F9 cells as well as CBA spermatogonia express a hitherto unknown functional analog of H-2 antigens. This argument would imply that, as with alloreactive CTL, one is dealing with CTL reactive to this postulated H-2 analog. It is also

TABLE 3. Specificity of Anti-F9 Immune CTL

Effector cells[a]	% Specific lysis of target cells[b]					
	F9	PCC$_3$	PCC$_4$	H-2^{bc} blasts	PYS	H-2^d blasts
CBA anti-F9	34	20	ND	4	2	5
CBA anti-H-2^{bc}	3	8	17	52	3	9

[a] Effector cells were induced *in vitro* (see legend to Table 1).
[b] The target cells F9, PCC$_3$, PCC$_4$ are F9 antigen-positive (10), the LPS blasts H-2^{bc}, H-2^d, and the PYS (10) target cells are F9 antigen-negative. The data refer to a ratio of attacker cells to target cells of 50 : 1. Background lysis of the target cells during the 5 hr cytotoxicity assay was less than 35%.

TABLE 4. Cytolytic Activity of CBA Anti-F9 Immune CTL for
CBA Derived Spermatogonia

Effector cells	% Specific lysis (E : T = 50 : 1)		
	F9	CBA spermatogonia[a]	H-2^k-blasts[a]
CBA anti-F9[b]	57	46	3
BALB/c anti-CBA[c]	ND	5	67

[a] CBA spermatongonia were prepared from 4-6 days old CBA mice by gentle trypsinizatin of cut pieces of the removed gonades. H-2^k blasts were prepared by culturing CBA-splenic lymphocytes in the presence of LPS over 72 hr as described (4).
[b] CBA anti-F9 immune CTL were induced *in vitro* as described.
[c] BALB/c anti-CBA immune CTL were induced in a primary MLC as described (14).

conceivable that these CTL are restricted by the postulated H-2 analog but specific for a "foreign" antigen, for example, foetal calf serum constituents. In any case, the experiments indicate that it is possible to induce functional CTL with H-2-negative stimulators. In terms of the two basic models for T cell recognition, dual recognition versus modified self (4), the results imply that the lytic activity of CTL can also take place in the absence of H-2 self-markers.

Acknowledgments

We are grateful to Drs. F. Jacob and H. Jakob, Paris, for supplying us with F9 cells. Dr. R. Kemler, Paris, kindly provided the mouse 129 anti-F9 serum used.

This work was supported by the SFB 107 (Mainz), BNRS, and Inserm (France).

References

1. Shearer, G. M., Rehm, G. T., and Garbarino, C. A. *J. Exp. Med.* **141**, 1348, 1975.
2. Forman, J. *J. Exp. Med.* **142**, 403, 1975.
3. Gordon, R. D., Simpson, E. and Samelson, L. E. *J. Exp. Med.* **142**, 1108, 1975.
4. Zinkernagel, R. M., and Doherty, P. C. *J. Exp. Med.* **141**, 1427, 1975.
5. Miller, J. F. A. P., Vadas, M. A., Whitelaw, A., and Gamble, J. *Proc. Natl. Acad. Sci. USA* **72**, 5095, 1975.
6. Erb, P., and Feldman, M. *J. Exp. Med.* **142**, 460, 1975.
7. Bach, F. H., Widmer, B. M., Bach, L. M., and Klein, J. *J. Exp. Med.* **136**, 1430, 1972.
8. Forman, J., and Vitetta, E. S. *Proc. Natl. Acad. Sci. USA* **72**, 3661, 1975.
9. Zinkernagel, R. F., and Oldstone, M. B. *Proc. Natl. Acad. Sci. USA* **73**, 3666, 1976.
10. Jacob, F. *Immunol. Rev.* **33**, 3, 1977.

11. Starzinski-Powitz, A., Pfizenmaier, K., Röllinghoff, M. and Wagner, H. *Eur. J. Immunol.* **6**, 799, 1976.
12. Pfizenmaier, K., Starzinski-Powitz, A., Wagner, H. and Röllinghoff, *Z. Immun.-Forsch.* **153** 268, 1977.
13. Wagner, H., Starzinski-Powitz, A., Röllinghoff, M., Golstein, P. and Jakob, H. *J. Exp. Med.* **147**, 251, 1978.
14. Wagner, H., Goetze, D., Ptschelinzew, L. and Röllinghoff, M. *J. Exp. Med.* **142**, 1477, 1975.
15. Klein, J., Geib, R., Chiang, C. and Hauptfeld, V. *J. Exp. Med.* **143**, 1439, 1976.
16. Kemler, R., Babinet, C., Condamine, H., Gachelin, G., Guenet, J. L. and Jacob, F. *Proc. Natl. Acad. Sci. USA* **73**, 4080, 1976.

CELL-MEDIATED IMMUNITY AGAINST AVIAN VIRUS-INDUCED TUMOR CELLS

Masanori Hayami, Jagodina Ignjatovich, Bernhard Fleischer,
Johannes F. Steffen, Helga Rübsamen, and Heinz Bauer

Extensive studies have been devoted to the surface of the transformed cell as a primary target of the host antitumor immune defense. The avian RNA tumor viruses are particularly useful for these investigations since this group includes avian sarcoma viruses (ASV) able to transform embryonic fibroblasts *in vitro*, as well as avian leukosis viruses (ALV), which only replicate in fibroblasts without transforming them [for review see Bauer (1)]. ASV and ALV are divided into several subgroups according to differences in their envelope glycoprotein, gp85, which defines the host range for certain avian cells and displays subgroup-specific and group-specific antigenic determinants (2). It was expected that transformation-specific events on the cell surface could be recognized using these viruses. To this end comparative experiments were performed with ASV- and ALV-infected targets.

It has been shown that spleen cells from Japanese quails, which have recovered from an ASV-induced tumor, exert a strong cytotoxic effect against allogeneic and autochthonous ASV-transformed quail fibroblasts (3). This cell-mediated cytotoxicity (CMC) is detected in an 18-hr visual cytotoxic assay that has been previously described (4). Target cells in this cytotoxic reaction were as follows: (a) Quail embryo cells (QEC) infected *in vitro* by ASV and ALV of various subgroups; (b) an established quail cell line (approximately 400 passages), designated R(-):Q, consisting of QEC transformed *in vitro* by the Bryan strain of Rous sarcoma virus (5); the cells are deletion mutants for the viral envelope glycoprotein (6); and (c) an established quail cell line, designated MC3-5:Q, derived from a methylcholanthrene-induced tumor.

TABLE 1. Blocking of CMC by Several Cell Extracts[a]

Target cells transformed by	Effector Cells induced by	Cell extracts	Blocking of CMC (%)
ASV-A	ASV-C	ASV-A transformed	95
ASV-A	ASV-C	ALV-A infected	39
ASV-A	ASV-C	R(-):Q	60
ASV-A	ASV-C	MC3-5:Q	56
ASV-A	ASV-C	nQEC	1

[a] Target cells (100-500/well) were seeded into Falcon Microtest I plates. After over-night incubation the medium and unattached cells were removed and spleen cells that had been preincubated (45 min, 37°C) with cellular extracts at a protein concentration of 100 μg/ml were added at an effector target cell ratio of 100-200:1. After 18 hr the spleen cells and detached target cells were removed by washing and the remaining cells stained and counted. Cytotoxicity was calculated as % destruction by immune effector cells compared with normal spleen cells applied in an identical way.

The aim of these experiments is to identify tumor-associated antigens and to elucidate their role in the interaction of the host's immune system with the tumor cell. This knowledge should lead to characterization of effector mech-anisms of antitumor immunity *in vitro* as well as *in vivo*, and to identification of discrete events in the host-tumor relationship.

RESULTS

As described in detail (4), several target antigens could be demonstrated by testing spleen cells from ALV- or ASV-injected quails. The former are able to destroy QEC infected *in vitro* by ASV or ALV of the same subgroup only. In contrast, spleen cells from quails immunized by ASV are highly cytotoxic for QEC transformed by a sarcoma virus of the same and of other subgroups; the destruction of ALV-infected QEC is subgroup-specific. R(-):Q cells and the chemically transformed MC3-5:Q cells are both destroyed to the same extent as are ASV-transformed cells. This destruction is presumably directed against anti-gens that are specific for transformation and different from viral structural proteins. A weak cytotoxic effect could also be observed on uninfected primary QEC but not on embryo cells of high-passage number.

The authors have recently found that pretreatment of the effector cells with extracts from various target cells block the cytotoxic reaction (7, 8). Table 1 gives an example of an experiment where the cytotoxicity is blocked completely by cell extracts from transformed cells. The transforming virus was the one that induced the immune cells *in vivo* (ASV-subgroup A). Partial blocking was obtained with extracts from ALV-infected cells as well as from R(-):Q and MC3-5:Q cells. This partial blocking is a qualitative effect as even high concen-

TABLE 2. Distribution of Target Antigens on Transformed and Nontransformed QEC

Target cells	s-gp85	g-gp85	EAg	TSSA
ASV-transformed QEC	+	+	+	+
ALV-infected QEC	+	−	−	−
R(-):Q	−	−	+	+
MC3-5:Q	−	−	+	−
nQEC of low passage (1-3)	−	−	(+)	−
nQEC of higher passage	−	−	−	−

trations of extract did not block completely. However, mixtures of extracts that by themselves caused partial blocking had additive effects.

Extracts from MC3-5:Q only partially block CMC against R(-):Q, whereas R(-):Q extracts completely block the reaction against MC3-5:Q. The cells transformed by the defective sarcoma virus seem to express at least one antigen in addition to those expressed on the chemically transformed MC3-5:Q. These and other experiments (8) performed in a chessboard pattern enabled us to distinguish four different target antigens. Two of these are antigenic determinants of the major viral envelope glycoprotein, one subgroup-specific (s-gp85), the other group-specific (g-gp85), and expressed only at the surface of transformed cells. As with g-gp85, two additional antigens are expressed specifically on transformed cells. One of these, EAg, is probably of embryonic origin, shared by ASV and methylcholanthrene transformed cells, and found on some embryonic cells of low-passage number. The other one, not a constituent of the virion itself but present on virus-transformed cells, is tentatively called transformation-specific surface antigen (TSSA). Table 2 summarizes the conclusions drawn from these experiments.

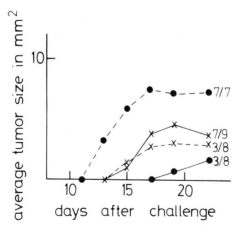

FIG. 1. Growth rate and incidence of tumor formation after ASV inoculation of Japanese quails immunized with extracts from R(-):Q (●) or nQEC (x). The animals were immunized six times at two-week intervals intravenously (——) or intramuscularly with adjuvant into the breast muscle (– – –). They were challenged with ASV-A two weeks after the last injection.

Extracts similar to those used for the *in vitro* experiments but prepared with nonionic detergent P40 or Triton X-100 have been used for *in vivo* protection experiments in Japanese quails after partial purification on *lens culinaris* lectin columns. Interestingly, it was found that the protective effect of immunization was strongly dependent on the route of administration. Intravenous injection decreased the incidence and the growth rate of tumors after subsequent virum inoculation, whereas intramuscular injection had an enhancing effect (Fig. 1) (9).

This phenomenon could be explained in part by the finding that spleen cells from quails immunized iv exerted a strong cytotoxic effect against cells carrying the immunizing antigens, whereas spleen cells from im injected animals showed no cytotoxicity whatsoever. Furthermore, spleen cells from im immunized quails were able to suppress the cytotoxicity of spleen cells from iv immunized donors (Table 3). Their admixture in a 1:1 ratio led to complete suppression of cytotoxicity. This was not due to steric hindrance or dilution of the immune cells, since spleen cells from nonimmunized animals (NQSC) had no such suppressing effect. As is evident in Table 3, the suppressive activity is dose dependent: With increasing ratio of iv:im immune spleen cells the cytotoxicity increased.

CONCLUDING STATEMENT

An intriguing problem that complicates the application of immunological methods to tumor prevention or therapy is the phenomenon of tumor enhancement under certain conditions. Various explanations have been developed but mechanisms remain obscure.

The system reported allows the reproducible induction of either tumor immunity or tumor enhancement in birds. It is shown that tumor enhancement is in part due to the development of suppressor cells. Since several tumor-

TABLE 3. Suppression of CMC by Spleen Cells from Animals Injected with Cellular Extracts

Material and route of injection into effector cell donors		% CMC[a]
ASV-A extract	iv	61[b]
ASV-A extract	im	0
ASV-A extract	iv + im (1 + 1)	0
ASV-A extract	iv + im (2 + 1)	14
ASV-A extract	iv + im (4 + 1)	34
ASV-A extract	iv + NQSC (1 + 1)	52
nQEC extract		0

[a] Target cells: ASV-B transformed
[b] Effector : target cell ratio = 100:1

associated cell surface antigens of different origins are distinguishable in this system, further analysis of their biochemical structure and function may elucidate their role in the various facets of immune responses.

Acknowledgments

This work has been supported by the Sonderforschungsbereich 47 of the Deutsche Forschungsgemeinschaft.

References

1. Bauer, H., *Adv. Cancer Res*. **20**, 275, 1974.
2. Rohrschneider, L. R., Bauer, H., and Bolognesi, D. P. *Virology* **67**, 234, 1975.
3. Hayami, M., Hellstr:
3. Hayami, M., Hellström, I., Hellström, K. E., and Yamanouchi, K. *Int. J. Cancer*, **10**, 507, 1972.
4. Hayami, M., Ignatovic, J., and Bauer, H. *Int. J. Cancer* **20**, 729, 1977.
5. Friis, R. R. *Virology* **50**, 701, 1972.
6. Halpern, M. S., Bolognesi, D. P., and Friis, R. R. *J. Virol.* **18**, 504, 1976.
7. Bauer, H., Ignjatovic, J., Rubsamen, H., and Hayami, M. *Med. Microbiol. Immunol.* **164**, 197, 1977.
8. Ignjatovic, J., Rubsamen, H., Hayami, M., and Bauer, H. *J. Immunol.* **120**, 1663, 1978.
9. Bauer, H., Hayami, M., Ignjatovic, J., Rubsamen, H., Graf, Th., and Friis, R. R. *In* "Three Days on Avian RNS Tumor Viruses" (S. Barlati, and C. de Guili, eds.). Pavia, Italy, 1979.

ALTERNATIVE ROUTES OF ENTRY FOR CELL SURFACE ANTIGENS INTO THE IMMUNE SYSTEM

E. Clark, P. Lake, N. A. Mitchison, and I. Nakashima

It has long been known that antigens presented on cell surfaces may enter the immune system through two alternative routes. One is the direct route: A foreign cell may encounter a lymphocyte directly, so that its antigens have a chance to bind directly to the receptors of the lymphocyte. The other is indirect: The foreign cell may break up, thus permitting its antigens to fix onto the surface of host macrophages, where they again have the chance to bind to lymphocyte receptors. These alternatives are illustrated in Fig. 1. Other possibilities seem unlikely, such as that antigens are shed and then bind to lymphocytes without the intervention of a host cell surface, because increasing evidence suggests that macrophages can only recognize antigens and lectins when these are presented on a cell surface (1).

The concept of alternative routes of entry stems from studies of tissue allografts, in which there was much discussion of the route by which antigens of the allograft reached host lymphocytes (2). One possibility was that host lymphocytes visited the graft, entering via the blood stream and leaving in a sensitized condition. Alternatively, antigens were shed from the graft, and passed via afferent lymphatics to the draining lymph node where they encountered host lymphocytes. The fate of antigens introduced into the body by skin painting with agents such as picryl chloride was discussed in very similar terms (3). These discussions were inconclusive, mainly because of the lack of techniques appropriate for resolving the problem in any general way. At best the proponents of one view could point to circumstances where anything other than a direct encounter between graft antigens and responder lymphocytes could be excluded on the grounds of speed of response; for example, during the sensitization of lympho-

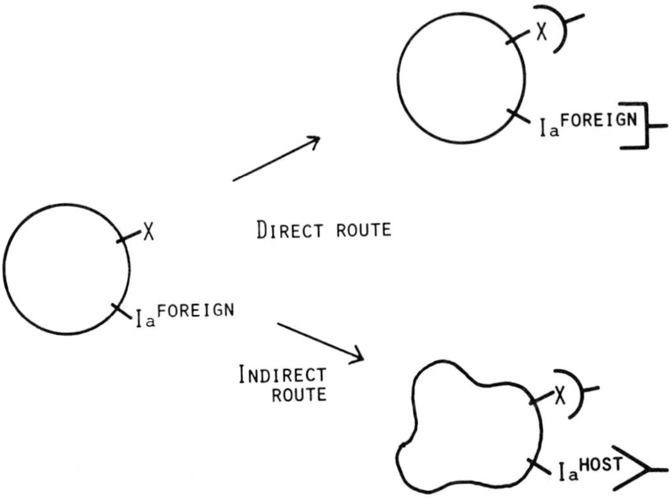

FIG. 1.   Alternative routes of antigen entry into the immune system.

cytes by passage through an isolated kidney (4). Proponents of the other thesis could argue that under other circumstances lymphocytes proved unable to respond to antigens presented on foreign cells without the intervention of macrophages; for example, during sensitization *in vitro* to the antigens of foreign erythrocytes (5).

Interest in this question revived quite recently in connection with the cytotoxic T-cell (T_c) response to minor transplantation antigens. It was noticed that under certain circumstances, T_c developed which displayed unexpected major histocompatibility complex (MHC) restrictions (6-8). It became clear that these apparent exceptions to the Zinkernagel-Doherty rules could be explained best by invoking a mechanism of host processing of the foreign minor antigens. These, then, are examples of what we now term the indirect route. As yet the circumstances that dictate which route will be used have not been systematically studied.

The purpose of the present chapter is to discuss these circumstances. It draws attention to the possibility that control of access by the indirect route may be an important mechanism of self-tolerance, and that breaches of this control may lead to autoimmune disease. It discusses experimental procedures that can be used to evaluate the contributions of the alternatives. It focuses attention on helper T cells (T_h) rather than T_c cells, partly because we are now engaged in a study of these cells, and partly to show that one can now ask the same kind of question of either type of cell.

Returning to Fig. 1, the crucial point is that under normal circumstances, a cell surface alloantigen entering the immune system via the direct route will encounter responsive lymphocytes in association with foreign Ia antigens (interesting combinations of mice can be selected in which donor and host share Ia antigens, but these are exceptions to the general rule). The Zinkernagel-Doherty rules state that T_h cells recognize foreign antigens only in association with self-Ia, the one exception being foreign Ia itself. The consequence of this is that T_h cells should be unable to recognize foreign alloantigens via the direct route, and must therefore avail themselves of the indirect route. Insofar as the indirect route operates more slowly or less efficiently than the direct route, T_h cells responding to antigens other than foreign Ia should develop relatively slowly and poorly during the immune response, in comparison with T_h cells responding to foreign Ia.

How securely established is the Zinkernagel-Doherty rule just stated? The evidence is that (i) in mixed lymphocyte cultures, cells of T_h type, as judged by helper activity and Ly-1 phenotype, are generated in response to Ia rather than non-Ia MHC antigens (9); and (ii) during *in vivo* sensitization to entirely foreign antigens, T_h cells are generated with genetically determined restrictions that map in the I region (the best data are those of J. Sprent, unpublished). One important piece of information is lacking, namely, whether non-MHC alloantigens behave toward T_h cells like Ia or non-Ia MHC antigens. If they behave like Ia antigens—as they may in fact do—the rest of the argument as developed here is in trouble.

LATENT HELP

What started us thinking about alternative routes of presentation was a study of helper relationships between separate molecules on the cell surface. We noticed that certain molecules, while well able to elicit a secondary response on their own, were unable to elicit a primary response without help from other molecules. This study is still imcomplete, but provisional data of our own and others are summarized in Table 1. This table lists pairs of donors, which for the hosts shown present either a single antigenic molecule, namely, Thy-1, H-2K, or H-2D, or a multiple combination of molecules including the one presented singly by the other donor. Thus, for example, cells taken from AKR/Cum donors into AKR/Jax hosts present the antigen Thy-1.2 alone, while cells taken from CBA donors into the same hosts present both Thy-1.2 and also several non-H-2 antigens. The classification of an antigen as single is, of course, a simplification.

At any rate, the antigens listed as single all either fail to generate an immune response under conventional conditions of immunization, or do so relatively inefficiently in comparison with the multiple combinations. This is not so, however, for the responses elicited in cells previously immunized by the multiple combinations: Here the single antigens are approximately as effective as the multiple combinations. Furthermore, as first shown by Wernet and Lilly (10), multiple immunizations with a single antigen may actually inhibit part of the subsequent response to the same antigen in multiple combination.

A possible explanation was that the secondary responses had become thymus-

TABLE 1. Latent Help System[a]

	1^0	2^0	Suppression
Thy-1.2 + H-minors	+	+	
Thy-1.2	−	+	?
(CAB, AKR/Cum→AKR/Jax)			
kkkkkdddb	+	?	
bbbbbbbbb	−	?	+
(B10.A(2R), B10→B10.A(5R))			
dddddddd	+	+	
bbbkkdddd	±	+	−
(B10.D2, B10.A(5R)→B10)			
bbbbbbbbb	+	?	
kkkkkdddb	−	?	?
(B10, B10.A(2R)→B10.A)			

[a] Each of the four triple-strain combinations comprises (i) a donor strain, differing from the host either by several minor antigens or by I-A and I-B; (ii) a donor strain not so differing; and (iii) the host strain. Antigens by which donor and host differ are underlined. For the last three strain combinations, the following MHC alleles are listed: K, I-A, I-B, I-J, I-E, I-C, S, G, D. 1^0 responses are to one or more injections of donor cells, 2^0 are adoptive secondary responses. Responses are IgM to Thy-1 (plaques or cytotoxic titers), IgG to MLC (cytotoxic titers) (see refs. 10 and 11).

independent. This has been eliminated by experiments in which adoptively transferred cells were treated with anti-Thy-1 plus complement. The conclusion follows that antigens which by themselves cannot elicit T_h cells efficiently can do so during the course of immunizations with combinations of antigens. Help of this type has been termed "latent" (12).

THE RATES OF GENERATION OF K-SPECIFIC AND Ia-SPECIFIC HELP COMPARED

If one wishes to measure the rate of generation of helper activity, experiments of the kind outlined in Table 1 are not satisfactory because they involve immunization of the T- and B-cell compartments simultaneously. This difficulty can be avoided by the experimental design illustrated in Fig. 2. The B-cell compartment is immunized in one group of mice against the antigen H-2D^c, by injecting cells from ccc donors into bbb hosts (we use an abbreviated notation for the MHC, in which bbb denotes H-2K^b, I-A^b, and I-B^b). In other groups of bbb mice, the T-cell compartment is immunized against aaa cells, using different numbers of injections in order to investigate the rate of response. The activity in this compartment can then be assessed separately against either K^a or I^a (or, in practice, because of the limited number of strains available, against K^a or I^a) by choosing

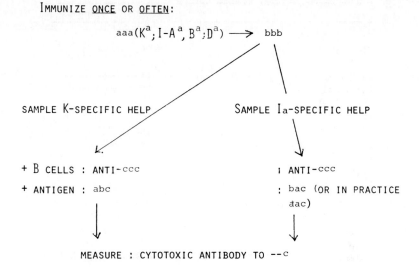

IMMUNIZE <u>ONCE</u> OR <u>OFTEN</u>:

$$aaa(K^a; I-A^a, B^a; D^a) \longrightarrow bbb$$

SAMPLE K-SPECIFIC HELP   SAMPLE Ia-SPECIFIC HELP

+ B CELLS : ANTI-ccc   : ANTI-ccc

+ ANTIGEN : abc   : bac (OR IN PRACTICE
                                                 aac)

MEASURE : CYTOTOXIC ANTIBODY TO --c

FIG. 2.   Schema for separate immunization of B- and T-cell compartments.

a suitable cell as immunogen in adoptive transfer, as shown in the figure. Thus, the rate of increase of activity against K and Ia can be compared.

Using this design, comparisons have been made in the generation of T_h cells against H-2^s and H-2^q (12). The results obtained in either case are much the same: maximum activity against Ia develops early, a single injection may be sufficient, whereas activity against K develops more slowly but eventually attains a similar plateau level. These findings thus provide further evidence of latent help, this time directed at H-2K^s or H-2K^q.

Experiments of this type are open to more than one interpretation. Thus, for example, K (or possibly an Ia-like antigen closely linked to K) may attract help slowly simply because it is rather an inefficient immunogen T_h-wise. The important point is that the results are compatible with the hypothesis formulated that foreign K-reactive T_h cells cannot recognize K in association with foreign Ia, and therefore cannot utilize the direct route of presentation. They can recognize K in association with self-Ia, the combination produced by the indirect route, but this occurs relatively slowly.

The hypothesis is now open to more direct test, perhaps using antigen-pulsed macrophages, or perhaps using abnormal Ia-association produced by cell fusion.

MHC-RESTRICTED T_h CELLS

MHC-restricted T_h cells have already been mentioned, in connection with the unpublished work of J. Sprent using sheep red cells as antigen. The same ques-

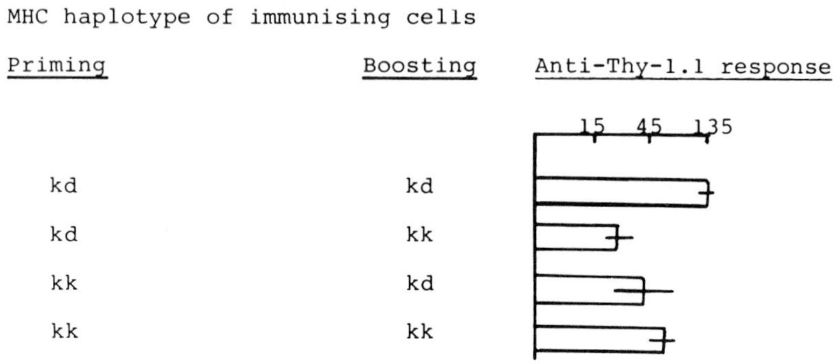

162   E. CLARK et al.

MHC haplotype of immunising cells

Priming   Boosting   Anti-Thy-1.1 response

                                                    15   45   135

       kd   kd

       kd   kk

       kk   kd

       kk   kk

FIG. 3. Dual recognition of Thy-1 and MHC. MHC restriction for the antibody response to the cell surface alloantigen Thy-1.1. Responses (kd x kk) F_1 hosts (A.Thy-1.1 x AKR) to priming and adoptive secondary boosting with kd (B10.A) or kk (CBA) cells; shown as excess over non-boosted response; cytotoxic antibody titres of groups of six mice, mean and s.d.

tion about MHC restrictions can be asked of cell surface alloantigens. An experiment aimed at this direction is illustrated in Fig. 3. In this experiment mice of genotype Thy-1.1, H-2$^{a/k}$ were immunized in separate groups with Thy-1.2, H-2^a and Thy-1.2, H-2^k cells. Cells from both groups were then boosted, after adoptive transfer, separately with cells of the two types used for immunization. The results show some evidence of MHC restriction, in that cells of the MHC type used for initial immunization proved somewhat more effective in boosting, than in the expected criss-cross pattern. Insofar as they do so, they provide evidence of the direct route of presentation operating. The combination of MHC types used was dictated by the strains available, and is unsatisfactory because H-2^a and H-2^k have the K-end of the MHC in common. We are breeding better strains for this purpose (e.g., Thy-1.1, H-2^d), and plan also to use the AK x L strains of B. Taylor. With their help, the contributions of the direct and indirect routes should be measurable in a rather direct way.

IMPLICATIONS FOR SELF-TOLERANCE AND AUTOIMMUNE DISEASE

Consider a differentiation product X, present on the surface of Ia-negative cells as shown in Fig. 4. According to the Zinkernagel-Doherty rule that we have invoked, X cannot elicit an autoimmune response unless it becomes associated with Ia as shown, e.g., on the surface of a macrophage. If this kind of inappropriate association never occurs, the immune system need not become actively tolerant of X. Presumably, therefore, any barrier to inappropriate associations will have survival value. It will develop in evolution not only directly as a barrier

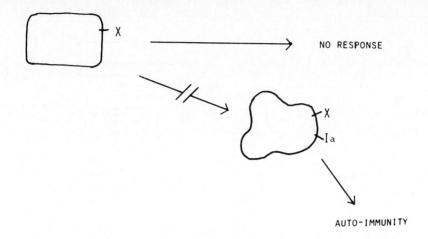

FIG. 4.   Schema for the generation of tolerance to self.

against autoimmune disease, but also indirectly as a mechanism whereby the load imposed on the immune system by active tolerance mechanisms can be reduced. Whether it is important or not we do not know; but its attractions as a first line of defence in self-tolerance are clear.

References

1. Habu, S., and Raff, M. *Eur. J. Immunol.* **7**, 451, 1977.
2. Medawar, P. B. *Proc. R. Soc. B.* **149**, 145, 1958.
3. Macher, E., and Chase, M. W. *J. Exp. Med.* **129**, 103, 1969.
4. Stoker, S., and Gowans, J. L. *J. Exp. Med.* **122**, 347, 1965.
5. Ford, W. L., Gowans, J. L., and McCullagh, P. J. *In* "Ciba Foundation Symposium" (G. E. W. Wolstenholme and R. Porter, eds.), p. 58. Churchill, London, 1966.
6. Mitchison, N. A. "VII Immunopathology Symposium" (P. Miescher, ed.), p. 28. Schwabe and Co., Basel, 1977.
7. Gordon, R. D., Mathieson, B. J., Samelson, L. E., Boyse, E. A., and Simpson, E. *J. Exp. Med.* **144**, 810, 1976.
8. Bevan, M. J. *Cold Spring Harbor Symp. Quant. Biol.* **41**, 429, 1976.
9. Bach, F. H., Bach, M. L., Kuperman, O. J., Sollinger, H. W., and Sondel, P. M., *Cold Spring Harbor Symp. Quant. Biol.* **41**, 429, 1976.
10. Wernet, D., Shafron, H., and Lilly, F. *J. Exp. Med.* **144**, 654, 1976.
11. Douglas, T. C. "Annual Report of Basel Institute for Immunology," 1977.
12. Mitchison, N. A., and Lake, P. *In* "ICN-UCLA Symposium on Immune System," Park City, Utah, 1977 (E. Sercarz, L. A. Herzenberg, and C. F. Fox, eds.), p. 555, Academic Press, New York. 1978.

CELL-MEDIATED CYTOTOXICITY AGAINST
TNP-MODIFIED HUMAN LEUKOCYTES

Stephen Shaw and Gene M. Shearer

Covalent chemical modification of murine lymphocyte surfaces can render them "antigenic" to autologous lymphocytes (1). Investigation of this phenomenon has permitted analysis of the specificity requirements of T-cell recognition, and demonstrated two roles of the major histocompatibility complex (MHC): one coding for cell surface structures recognized in association with the modifying agent, and the other controlling levels of response in cell-mediated immunity (2). Responses of T lymphocytes to trinitrophenyl (TNP)-modified syngeneic cells (self-TNP) have been demonstrated both in assays of cell-mediated lympholysis (CML) (1) and of secondary proliferative response (3). Although these responses may be mediated by distinct T-cell populations, both are characterized by similar specificity requirements: the moiety recognized is the modifying agent (TNP) in association with self-determinants controlled by genes mapping in the MHC. Broader relevance of this MHC-restricted phenomenon is suggested by studies showing similar specificity requirements for CML in virally infected (4) and minor transplantation antigen-associated systems (5, 6). Extension of this technique of TNP modification to human studies would allow testing of the hypothesis that the human immune system resembles the murine system with respect to the nature of T-cell receptor specificity and its dependence on genetic control by the MHC.

GENERAL CHARACTERISTICS OF THE HUMAN
CYTOTOXIC RESPONSE TO SELF-TNP

Human peripheral blood lymphocytes cultured *in vitro* with TNP-modified autologous stimulator cells generate cytotoxic effectors that are capable of

165

killing TNP-modified autologous cells. Recently, Newman *et al.* (7) reported this observation using a culture system with nylon nonadherent responder cells. We have made a similar observation in a system using unseparated peripheral blood lymphocytes. The ability to generate such a response appears to be a common, if not universal, characteristic of normal donor cells, as we have observed no "nonresponders" out of more than 30 donors tested.

Some of the general characteristics of the cytotoxic response to TNP-modified autologous cells are illustrated in Table 1. Cells from two unrelated donors were sensitized in primary culture to stimulating allogeneic cells, autologous cells, or TNP-modified autologous cells. After eight days in primary culture, cells were harvested and assayed in a standard 4-hr chromium release assay on modified and unmodified cells from self and the unrelated donor. Conditions for human cell culture have been adapted from work of Mawas *et al.* (8) and Bach *et al.* (9) in human allogeneic systems; TNBS modification was as described by Shearer (10). The important modifications were the use of autologous plasma as a media supplement (which improved the reproducibility of the response) and the elimination of mitogen stimulation of the lymphocyte targets (which improved specificity).

Cytotoxic activity against TNP-modified autologous cells could be demonstrated in primary response with cells sensitized against either of two immunogens: TNP-modified autologous cells or allogeneic cells (Table 1). In the former case, where cytotoxicity was generated in a totally autologous system, the response was TNP-dependent both at the sensitization phase and the effector phase. That is, the activity of anti-self-TNP effectors could be detected only on TNP-modified targets (lines 2 and 5, no activity on unmodified A or B), and their generation required TNP presentation on the stimulating cell (line 1 vs. 2 and 4 vs. 5).

Although the activity of these anti-self TNP effectors is the subject of this discussion, it should be noted in passing that cytotoxic activity was (and is generally) observed when cells sensitized to allogeneic stimulators were assayed on TNP-modified (but not unmodified) autologous cells. The amount of activity observed was only 1-2% (based on comparison of lytic units, as is generally the case in this discussion) of that observed with the same effector on the specific allogeneic target and also less than 10% of that observed with effectors specifically sensitized to self-TNP. This lysis may have a trivial explanation such as low-affinity nonspecific interactions between TNP-modified cells and activated effector cells, as has been suggested for murine lymphocytes modified with periodate (1). Alternatively, the lysis may reflect specific cross reactivity between certain allogeneic determinants and neoantigens on TNP-modified autologous cells (11). The latter hypothesis has been suggested and approached in murine studies, but definitive answers in both species await specificity analysis of the progeny of individual cytotoxic cell precursors (12).

We predicted that the human response to self-TNP, like the murine response, would depend not only on recognition of TNP but also on cell surface com-

TABLE 1. Cell-Mediated Cytotoxicity against TNP-Modified Human Lymphocytes[a]

		% Specific lysis assayed on targets designated at effector to target cell ratios indicated											
		A			ATNP			B			BTNP		
Responder	Stimulator	40	10	2.5	40	10	2.5	40	10	2.5	40	10	2.5
1 A	A	0.2	0.8	—	-0.9	—	—	-1.6	-1.9	-0.5	—	1.5	-0.2
2 A	ATNP	-1.9	0.8	0.4	22.3	11.1	5.6	—	—	—	6.1	—	—
3 A	B	0.5	—	—	4.2	—	—	42.3	—	13.9	31.1	—	13.8
4 B	B	—	—	—	—	—	3.5	2.5	—	—	-2.8	—	—
5 B	BTNP	3.0	—	—	19.9	11.5	—	-2.0	—	—	21.7	8.0	3.7
6 B	A	61.6	—	36.5	51.0	—	36.8	-0.8	—	—	7.5	—	—

[a] Responding lymphocytes were cultured in the presence of irradiated stimulating cells, harvested on the eighth day and assayed on ^{51}Cr-labeled modified and unmodified target cells.

TABLE 2. Cold Target Inhibition of TNP-Dependent Cell-Mediated Cytotoxicity[a]

Effector from donor #	Target	Donor cell serotyping	% Lysis In presence of donor cells:								% Inhibition[b] In presence of donor cells:						
			—	2	3	1TNP	2TNP	3TNP	4TNP	5TNP	2	3	1TNP	2TNP	3TNP	4TNP	5TNP
1	1TNP	1,2,4,8,51	28.2	12.8	16.5	0.8	6.6[c]	14.6	12.9	12.7	55	41	97[c]	77[c]	48	54	55
2	2TNP	1,2,4,8,51	27.4	21.1	23.8	10.6[c]	2.0	24.8	21.7	22.6	23	14	61[c]	93	9	21	18
3	3TNP	2,3,7,12	32.4	17.6	18.2	15.5	11.7	1.3	7.4[c]	4.1[c]	46	44	52	64	96[c]	77[c]	87[c]
4	4TNP	2,3,7,12	25.8	17.2	15.8	7.9	10.4	10.6[c]	0.4	8.6[c]	33	39	69	60	59[c]	99	67[c]
5	5TNP	2,3,7,12	21.6	10.8	12.0	9.1	9.9	5.6[c]	6.6[c]	1.0	50	45	58	54	74[c]	69[c]	95

[a] Cells from each donor were sensitized in primary and secondary *in vitro* culture to self-TNP and assayed at a 5:1 effector-to-target cell ratio on TNP-modified autologous target cells. Data shown are for a blocker-to-target cell ratio of 60:1. Underline indicates blocking with cold target identical to labeled target.

[b] % Inhibition is calculated as (% lysis in presence of designated blocker)/(% lysis without blocker).

[c] Indicates combinations in which target cell and blocker are matched for all HLA-A and B antigens.

ponents coded for by MHC-linked genes. We have observed, as expected, a component of MHC-restricted self-specificity in the cytotoxic response to TNP, but this has been difficult to analyze in the human response. Our current understanding of the specificity can be summarized as follows:

(1) The cytotoxic activity is mediated largely, if not exclusively, by T cells.

(2) All the cytotoxic activity depends on recognition of TNP in association with cell surface determinants.

(3) Activity can be demonstrated against TNP in association with at least three broad classes of cell surface determinants. The classes of determinants are as follows:

(a) "Common determinants" that are widely shared (e.g., phenotype frequency in excess of 0.45) among unrelated Caucasian donors with no shared A and B locus determinants.

(b) "SD-associated determinants" that are highly polymorphic among Caucasians and are strongly associated with A and B locus determinants.

(c) "Non-SD-associated determinants" that are HLA-linked but not strongly associated with A and B locus determinants, and are relatively polymorphic among Caucasians.

This discussion elaborates on the experimental basis for these conclusions and on some of their implications.

The specificity requirements of TNP-dependent cell-mediated cytotoxicity was analyzed either by direct lysis or by cold target cell competition of lysis. In either case, a panel of donors was selected as dictated by the hypothesis to be tested. Effector cells against self-TNP were generated from each donor's peripheral blood leukocytes by *in vitro* primary and secondary sensitization. Effectors harvested from secondary rather than primary cultures were used because they were 20-100-fold more active; their specificity was not detectably changed by the restimulation. In assays of direct lysis, effectors from each donor were assayed at varying effector-to-target cell ratios on modified and unmodified target cells from each donor. Significant lysis on unmodified target cells was not observed; all targets referred to from this point on are those that have been TNP-modified.

Initial studies of specificity and cross reactivity in TNP-dependent CML were conducted with cells of unrelated donors who shared no serologically defined HLA-A and B locus determinants. Effectors were generated from three such donors and assayed on modified targets from each (Fig. 1). With each effector population, lysis was greatest on self-TNP (significantly different from unrelated targets in each case; $p < 0.01$ for five combinations and $p < 0.05$ for 1). However, significant activity was observed on all modified targets. At least 20% cross reactivity has been observed, not only in these donor combinations, but in each other combination in which anti-TNP effectors have been assayed on modified allogeneic targets (>100 tested). On the average, 45% of the lytic activity on self-TNP was demonstrable on unrelated modified cells.

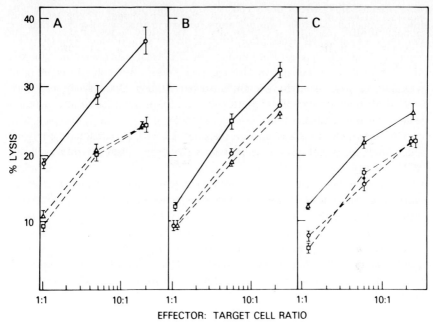

FIG. 1. Specificity of TNP-dependent cytotoxicity among three HLA-A and B unrelated donors. Cells from each donor were sensitized *in vitro* to self-TNP. Each panel represents such effector cells from one donor assayed on TNP-modified target cells from that donor (———) and from the other donors (– – –).

ALLOGENEIC CROSS REACTIVITY IN TNP-CML

The consistent high degree of cross reactivity had not been expected for a T-cell-mediated cytotoxic response. However, results of experiments to define cell type and effector mechanism have been consistent with an exclusively T-cell-mediated cytolysis. First, studies in collaboration with D. Nelson have established that most of the cytotoxic activity was unaffected by depletion of FcR+ cells, but was abolished by T-cell depletion. Second, cytotoxic activity was blocked rather than enhanced by addition to the assay of anti-TNP antibody capable of mediating antibody-dependent cellular cytotoxicity (ADCC). Third, cytotoxic activity was minimally reduced by addition of TNP lysine to the assay in concentrations that abolished anti-TNP ADCC. In each of these experimental manipulations, the activity on autologous and unrelated cells was similarly affected. The foregoing observations suggest that most of the cytotoxicity was mediated by T cells and ruled out ADCC as a predominant mechanism for the "specific" or "cross-reactive" cytotoxicity.

Since T-cell-mediated killing has been found to be largely self-specific in many studies with mouse cells, this broadly cross-reactive component of T-cell-mediated killing with human cells prompted further investigation. One possible explanation was that the effectors mediating this broadly cross-reactive lysis required only TNP recognition to activate the lytic process. The study referred to, in which lysis was not inhibited by TNP-lysine, suggested that this was not the case. Proof that exclusive TNP recognition was not involved was provided by studies of cross reactivity across xenogeneic barriers. Mouse effectors against self-TNP mediated some cross reactive lysis on other modified mouse cells but not on modified human cells. Conversely, human effectors against self-TNP cross reacted with modified human targets but not with modified mouse targets. The complete absence of xenogeneic cross reactivity suggested that exclusive TNP recognition was not sufficient for target cell lysis (since xenogeneic killing by T cells can be readily demonstrated when the appropriate specificities are recognized). Therefore, the cross reactivity observed within these species is most consistent with effector cell recognition of TNP in association with determinants shared within but not between species.

Among unrelated humans, average cross reactivity in TNP-dependent CML is approximately 45%. This finding indicates that, on the average, at least 45% of the cytotoxic activity is directed against self-determinants that are shared by at least 45% of donors. More accurate evaluation of the population frequency of these common determinants can be obtained by cold target-cell competition studies of the following experimental design: anti-self-TNP effectors from one donor assayed on modified targets from an unrelated donor and blocked by a panel of unrelated modified third-party cells. Preliminary studies of such design suggest that at least 30% of the activity is directed against determinants shared by most Caucasian donors.

HLA RESTRICTED ACTIVITY IN TNP-CML

Total cross reactivity of TNP-dependent cytotoxicity has been observed between certain pairs of siblings; other pairs are no more cross reactive than would be expected for unrelated individuals. Analysis of the correlation between such cross reactivity and sharing of HLA haplotypes was undertaken to determine if the increased cross reactivity could be attributed to determinants coded for by HLA-linked genes. Representative results are illustrated for the cytotoxic activity of effectors from one sibling on modified target cells from self and each of three siblings (Fig. 2). The amount of lysis varied between targets but generally paralleled the number of HLA haplotypes shared by the effector cells and target cells. Lysis was greatest on the HLA identical cells, less on the HLA-haploidentical cells, and least on HLA-unrelated cells. However, with these effectors, lysis on modified autologous cells was lower than on HLA-identical allogeneic cells. Similar observations have been repeatedly made and appear to be best explained

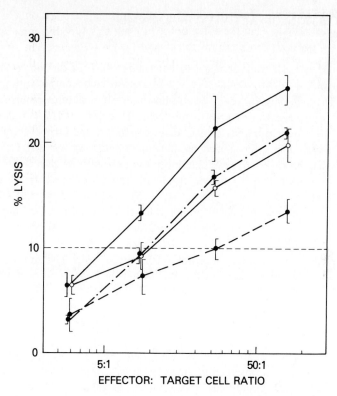

FIG. 2. Specificity of TNP-dependent cytotoxicity in a family study. Anti-self-TNP effectors from one sibling were assayed on self-TNP targets (○—○); or on TNP-modified targets sharing both HLA haplotypes (●—●), one HLA haplotype (●—·—·—●), or neither HLA haplotype (●———●).

in terms of differences between target cells in their susceptibility to TNP-dependent lysis. The complexities introduced by these differences in target cell lysability make it difficult to interpret raw data; however, variance analysis (13) corrects for such nonspecific effects and maximizes the information obtained from studies of direct lysis. Variance analysis of the results of direct lysis among all cell combinations from six siblings confirmed a strong association between the number of HLA haplotypes shared by siblings and their cross reactivity in TNP-CML. The activity of effectors on targets from HLA-unrelated siblings (mean cross reactivity 45%) was comparable to that seen among unrelated individuals. The activity of effectors on targets from HLA-haploidentical targets (mean 76%) was significantly less than on HLA-identical targets ($p < 0.01$) and greater than on HLA-unrelated targets ($p < 0.01$). The results, observed by direct lysis in this family study, were consistent with the hypothesis that the observed difference between lysis on self-TNP and on unrelated TNP-modified targets was due to recognition of TNP in association with determinants coded for by genes linked to HLA.

SD-ASSOCIATION OF ACTIVITY IN TNP-CML

Subsequent studies were designed to determine which of the loci within the HLA complex were responsible for the HLA-restricted component of cytotoxicity. The mapping of genes controlling the self structures recognized by murine effectors against self-TNP to the K and D regions of H-2 (1) raised the possibility that their putative human analogs (HLA-A and HLA-B loci) would code for determinants recognized by human effectors against self-TNP. This hypothesis has been tested by evaluating cross reactivity of cells from donors matched for HLA-A and B locus determinants. Studies of direct lysis among a panel of three donors of serotype 1, 24, 8, 51, and three donors of serotype 2, 3, 7, 12 demonstrated that matching individuals on the basis of those serologic markers was associated with increased cross reactivity between them in TNP-dependent cytotoxicity. These correlations suggested that sharing of all four A and B locus markers was not sufficient to make TNP-modified allogeneic cells totally cross reactive with self-TNP. That is, there were at least two components to the self-specificity; one due to A and B locus-associated determinants and another that was not. Recognition of the former was most apparent at high effector cell concentrations; but the latter dominated at low effector cell concentrations.

"Cold target competition" studies of the same effectors were performed to confirm and clarify the observations made by direct lysis. A variety of unlabeled cells (modified and unmodified) were assayed for their ability to inhibit the specific lysis of the effectors on TNP modified ^{51}Cr-labeled autologous targets (Table 2). For each combination of effectors and blocking cells, data have been expressed both in terms of % lysis and % inhibition of lysis (calculated as % lysis in the presence of designated blocker divided by % lysis without blocker). On the average, 39% reduction of lysis was observed in the presence of unmodified blocking cells. Since no interaction between such cells and TNP-sensitized effectors had been demonstrated by studies of direct lysis, this inhibition presumably represented nonspecific effects of large numbers of bystander cells Blocking was virtually complete (96% inhibition) when the ^{51}Cr-labeled targets and "cold" blockers were identical, as predicted for such "specific" inhibition. Blocking by TNP-modified unrelated cells (47% inhibition) was only slightly (and not significantly) more effective than by unmodified cells. On the average, 72% inhibition was observed in the presence of modified allogeneic blocking cells that were matched with the specific target for HLA-A and B locus antigens. Such blockers were significantly less effective than the modified autologous blockers ($p < 0.001$), but were significantly more effective than modified unrelated blockers ($p < 0.003$). These results confirmed the findings by direct lysis that each TNP-modified cell presented at least two kinds of polymorphic antigens; at least one that was associated with A or B locus antigen, and another that was not.

Although the foregoing analysis of blocking in terms of % lysis could identify distinct classes of antigenic determinants, it could not be used reliably to quantitate the proportion of cytotoxic activity against each class of determinants. More elaborate calculations suggest that activity against SD-associated determinants is several-fold greater than that against non-SD-associated determinants in most effector cell populations (although this may not be true of some effector cell populations). Activity against "common determinants" is not well demonstrated by such blocking studies; this is expected for reasons relating to the shallow slope of the effector cell titration curve, and is not inconsistent with the 45% average cross reactivity calculated from studies of direct lysis.

DISCUSSION

In any sensitization, effector cells are raised against a finite set of determinants. A fraction of the cytotoxic activity of the resulting effector cell preparation will be directed against each of these determinants. Studies of specificity of the cytotoxic response to TNP attempt to define which cell surface determinants are recognized in association with TNP and what the relative cytotoxic activity is against each of those determinants. The determinants can be defined in terms of the genetic locus that controls them, and in terms of their frequency in the donor population.

In murine studies of the anti-TNP response, at least two classes of determinants have been defined. One class is coded for by genes mapping in serologically defined regions of the H-2 complex (1). Most of the cytotoxic activity is directed against determinants of this type. Responses to another class of determinants, coded for by H-21 region genes, has been demonstrated recently (14). Cytotoxic activity against this class of determinants is estimated to represent 2-5 % of the total. There appears to be extensive polymorphism for both classes of determinants as assessed by the amount of cross reactivity among strains.

The classes of determinants recognized in the human cytotoxic response to self-TNP resemble to some degree those defined for the mouse. One such class is highly associated with serologically defined HLA-A and B locus antigens. A second class of determinants is not highly associated with those SD antigens but appears to be HLA-linked. These two classes of HLA-associated determinants are highly polymorphic among Caucasians. A third class of "common" determinants exists for which there is little, if any, polymorphism among the donors studied; these determinants are not shared between humans and mice. Genetic analysis of HLA-linkage and SD-association of these determinants will depend on identification of human polymorphism.

Analogies to one of these three components of the TNP-dependent CML have been reported in two other human cytotoxic responses to "modified-self" antigens. At this conference, McMichael has reported HLA-restricted cytotoxicity against influenza-infected human cells (15). The specificity analysis presented suggests that the self determinants recognized on influenza-infected cells are highly associated with the HLA-B locus (and to a lesser extent HLA-A locus) determinants. Recently Goulmy et al. (16) have reported in vivo generated cytotoxicity against a human sex-linked antigen. Killing by effectors from this individual was restricted to target cells that presented the sex-linked antigen in association with the appropriate HLA-A antigen (HLA-A2 that was shared by the effector cell and the specific immunizing cell). The cytotoxicity observed in these two systems resembles one of the three components detected in the TNP-dependent CML (i.e., the recognition of the "modifier" in association with cells bearing the relevant SD antigen).

Thus, it has been clearly demonstrated in human and mouse systems that self SD-associated determinants are frequently recognized together with the "modifying agents." Our findings suggest that at least two other classes of determinants can be recognized in association with TNP. First, non-SD-associated HLA-linked antigens can provide such recognition structures. These determinants may be analogous to the I-region structures recognized by mouse T-cells in association with TNP (14). Second, widely shared human antigens appear to be recognized in association with TNP. Such shared antigens could account for the more limited cross reactivity in TNP-dependent CML that has been observed between H-2 unrelated mouse strains (11).

Results from the human TNP-dependent CML suggest that a variety of self determinants can be recognized in association with TNP. It is unclear why the variety of self structures recognized in this system appear to be broader than in the other human systems that have been reported. Differences in the patterns of specificities observed in these human cytotoxic systems could be accounted for by any of at least three general mechanisms: (a) selective association of each modifying agent with different self structures; (b) antigen-specific genetic limitations of responder cell recognition (e.g., Ir genes); and (c) antigen-specific control mechanisms for regulation of the expansion of potentially responsive clones (e.g., suppression). One hypothesis would be that the broader range of determinants recognized in association with TNP could be due to the nonspecific nature of covalent coupling of TNP to cell surfaces. Alternatively, the relative high immunogenicity of the TNP moiety could circumvent certain genetic limitations in the response. It is also possible that the high concentration of TNP on cells during the in vitro sensitization may minimize the selective expansion of high-affinity clones. Regardless of the explanation of this complexity of the human response to self-TNP, the findings suggest that there may be more diversity than had been previously demonstrated in self structures.

References

1. Shearer, G. M., Rehn, T. G. and Schmitt-Verhulst, A.-M. Transplant. Rev. 29, 222, 1976.
2. Shearer, G. M. and Schmitt-Verhulst, A.-M. Adv. Immunol. 25, 55, 1977.

3. Schmitt-Verhulst, A.-M., Garbarino, C. A. and Shearer, G. M. *J. Immunol.* **118**, 1420, 1977.

4. Doherty, P. C., Blanden, R. V. and Zinkernagel, R. M. *Transplant. Rev.* **29**, 89, 1976.

5. Bevan, M. J. *Nature* **256**, 419, 1975.

6. Gordon, R. D., Simpson, E. and Samelson, L. E. *J. Exp. Med.* **142**, 1108, 1975.

7. Newman, W., Stoner, G. L. and Bloom, B. R. *Nature* **269**, 1977.

8. Mawas, C. E., Charmot, D. and Sasportes, M. *Immunogenetics* **2**, 449, 1975.

9. Bach, F. H. and Sheehey, M. J. *Tissue Antigens* **8**, 157, 1976.

10. Shearer, G. M. *Eur. J. Immunol.* **4**, 527, 1974.

11. Lemmonier, F., Burakoff, S. J., Germain, R. N. and Benacerraf, B. *Proc. Natl. Acad. Sci. USA* **74**, 1229, 1977.

12. Teh, H.-S., Phillips, R. A. and Miller, T. G. *J. Immunol.* **118**, 1057, 1977.

13. Takasugi, M. and Mickey, M. R. *J. Natl. Cancer Inst.* **57**, 255, 1976.

14. Wagner, H., Starzinski-Powitz, A. Jung, H. and Röllinghoff, M. *J. Immunol.* **119**, 1365, 1977.

15. McMichael, A. J., Ting, A. Zweerink, H. J. and Askonas, B. A. *Nature* **270**, 524, 1977.

16. Goulmy, E., Termijtehen, A., Bradley, B. A. and van Rood, J. J. *Nature* **266**, 544, 1977.

HLA RESTRICTION OF CYTOTOXICITY AGAINST INFLUENZA-INFECTED HUMAN CELLS

A. J. McMichael

It has been shown in mice that T-cell-mediated lysis directed against virus-infected target cells only occur when effector and target cells share H-2K or H-2D antigens (1). This was originally shown for lymphocytic choriomeningitis virus by Zinkernagel and Doherty (2) and has since been shown for vaccinia, ectromelia, Friend, and influenza viruses (3-7). More recent work has shown that only particular H-2 alleles will interact with the Friend virus (8) and lymphoma-inducing viruses (9).

In man, HLA restriction also applies to cell-mediated lysis directed at the Y transplantation antigen (10), dinitro-phenylated cells (11) and influenza virus-infected cells (12). The experiments described here extend our previous results using influenza-infected cells as antigen and indicate that the cytotoxic effector cells show considerable specificity toward self.

METHODS

The experimental procedures used have been previously described (12). In brief, peripheral blood mononuclear cells were infected with influenza virus (influenza X31: A Hongkong/68 X Ao/PR8; H_3N_2), incubated for 4 hr and then added to autologuos lymphocytes. The cell mixture was incubated for 5-8 days and the effector cells generated were then tested for lytic activity on freshly prepared, infected, and ^{51}Cr-labeled target cells at a killer-to-target ratio of 50:1. Chromium release into the supernatant was measured at the end of 5 hr incubation. Percent specific lysis was calculated as

$$\frac{^{51}\text{Cr release} - \text{medium control}}{^{51}\text{Cr release by Triton} \times 100 - \text{medium control}} \times 100$$

Background chromium release by medium alone (medium control) was normally around 20% (range 9-26%) of that released by Triton X 100.

RESULTS AND DISCUSSION

Our previous results (12) have shown that cells from an individual homozygous for HLA-B7 and sensitized to influenza virus-infected autologous cells lysed only target cells that shared HLA-B7. The cytotoxic cells also showed specificity for the influenza virus-type A or B. Mouse cytotoxic T cells do not distinguish influenza virus haemagglutinin from neuraminidase subtypes, either in the sensitization phase or in the lytic phase (6). This may explain the relative ease with which cytotoxic cells were generated in man because it is probable that all individuals studied had been immunologically primed by natural infection with one or more of the cross-reacting influenza A viruses.

The results of the experiment shown in Table 1 extend these findings using sensitized cells from a second individual FW, which is homozygous for HLA-A1 and B8. Lymphoid cells from FW, sensitized *in vitro* to autologous target cells infected with influenza virus A(31), lysed only influenza-infected target cells from donors that shared HLA A1 or B8.

TABLE 1. Lysis of Influenza-Infected Target Cells by Sensitized Lymphocytes of Individual FW (HLA: 1, 8, W3 / 1, 8, −)

Target cells	HLA[a]	Percent Specific Lysis[b] (%)
FW	1,8,W3/1,8−	41.7
JF	1,8,W3/W24,W35,−	33.5
JP	1,8,W3/W24,W35,−	32.5
DW	1,W32,W35,13,−,−	35.3
RG	1,3,14,15,−,−	33.0
RD	2,2,8,12,W3,−	32.2
CH	9,11,7,39,W3,W2	7.2
AT	2,11,40,22,−,−	10.2

[a] HLA antigens are shown either as genotype (A,B,D/A,B,D) or as phenotype A, A, B, B, D, D,. Shared antigens are underlined.
[b] 5 hour ^{51}Cr-release assay at killer : target ratio of 50 : 1.

Sensitized FW cells have been tested on 20 influenza-infected target cells. None of the six that did not share HLA antigens gave more than 10.2% specific lysis.

One target that shared only HLA-DW3 was not lysed, which is compatible with mouse data because HLA-D is probably equivalent to the H-2I region. Of 13 target cells that shared HLA A or B antigens, 12 were lysed by FW. The one exception was a target cell that only gave 6.1% lysis in spite of sharing HLA-A1.

Table 2 summarizes results with other sensitized cell donors. These results are expressed as "relative lysis" for ease of comparison. Each sensitized cell displayed a bimodal pattern of lysis with a group of target cells, and accordingly, one group of target cells was killed (relative lysis >58%), whereas another group was not as susceptible to lysis (relative lysis 0-32%). The lysed targets all shared HLA-A or B antigens with the killer cell. Targets that shared no HLA antigens were not killed, but each of the targets shown in Table 2 was lysed by at least one effector cell type. Thus, no target was intrinsically resistant to infection. In view of the reproducible infection in targets CW, FW, AM, the low levels of killing seen when HLA antigens were not shared probably reflect the failure of target cell recognition by effector cells.

The results are consistent with the phenomenon first described by Zinkernagel and Doherty (2) and may be explained by either of the models they proposed: dual recognition or recognition of an HLA-virus complex (altered self). There are

TABLE 2. Relative Lysis of Influenza-Infected Target Cells by Lymphocytes of Sensitized Donors[a]

Targets and HLA		Lymphocyte donors				
		CW 9,7,W2 9,7,W1	FW 1,8,W3 1,8,–	CH 9,7,W2 11,W16,–	JR 2,12,– 2,27,–	AM 2,21,– 32,W40,–
PGF[b]	3,7,W2/3,7,W2	101	–	160	21	8
JD	3,7,W2/2,7,W2	100	15	–	–	–
CW	9,7,W2/9,7,W2	100	22	–	–	–
FW	1,8,W3/1,8,–	22	100	–	–	–
CH	9,11,7,W16,W2,–	–	3	100	–	–
JR	2,12,–/2,27	13	–	0	100	23
AM	2,21–/32,W40,–	32	–	–	–	100
PM	2,7,W2/	85	–	60	30	6
AT[c]	2,11,W40,–,–	18	25	0	–	0
PF	3,7,W2/24,8,–	58	–	–	–	0
OD	2,7,W2/2,13,–	96	–	–	–	30
DH	2,7–/3,7,W2	84	–	64	22	–
JM	2,2,27,14	0	–	0	76	–

[a] Relative lysis: Specific lysis expressed as a percentage of the lysis seen with autologous infected targets. Specific lysis values from which those calculations were made were: 42 and 32% for CW, 42% for FW, 24% for CH, 33% for JR and 32% for AM. Negative values are expressed as 0, – = not tested.

[b] PGF is a lymphoblastoid cell line.

[c] Infected AT has been lysed by autologous sensitized cells.

some exceptions to the HLA restriction that deserve comment. All these have been in the direction of a greater specificity for self. Nonspecific cytolysis of all infected target cells, which has been reported for measles (13) and vaccinia-infected human cells (14), was not encountered. Of the effector cells tested, AM showed the utmost specificity as it failed to lyse any HLA-matched target cell. JR cells did not kill target cells that shared only HLA-A2; FW did not lyse one target that shared A1, and CW did not lyse one target that shared A9. These exceptions all occur in combinations with shared A locus antigens. It is unlikely that this represents an A locus effect simply because FW lyses two targets that share only HLA-A1.

Three other explanations for these exceptions seem possible. There may be HLA antigen subgroups that are serologically silent and killer target interaction between the two subgroups are not discernible. This would be comparable to the results described by Zinkernagel (15) with H-2 mutants.

A second possibility is that lysis is only seen when killer and target share an HLA haplotype. The instances in which lysis was seen with apparent single antigen sharing may reflect recent recombinations, where partial haplotype sharing did still occur. On the other hand, an individual such as AM could have a group of HLA antigens that are in linkage equilibrium such that unrelated target cells sharing an HLA haplotype would be hard to find and thus AM cells would display extreme specificity toward self. This explanation implies that other as yet unidentified HLA products, such as Ir genes (16), might play roles in target cell recognition.

A third possibility is that certain HLA antigens fail to interact with the influenza virus. This would be comparable to the finding of Blank and Lilly (8) according to which Friend virus only interacts with H-2D^b. HLA-A2 might fail to interact with the influenza virus because neither JR- nor AM-sensitized cells would kill infected targets that shared only HLA-A2. If this interpretation is correct, the effect would be antigen specific. Goulmy et al. (10) and Dickmeiss et al. (11) have shown that in the cytotoxic cell recognition of the H-Y antigen and dinitrophenylated cell surface, respectively, HLA-A2 sharing was particularly effective.

These explanations for the failure of sensitized cytotoxic cells to lyse particular target cells despite their HLA sharing are currently being tested. The results may be of consequence for the mechanisms of interaction between influenza viruses and the HLA system.

Acknowledgments

I thank Ms. Jenny Pilch for expert technical assistance, Dr. A. Ting for HLA typing the blood donors, Dr. J. A. Skehel for the gift of influenza virus, and Professor P. J. Morris and Dr. B. A. Askonas for advice and encouragement. This work was supported by a senior research grant to Professor Morris.

References

1. Doherty, P. C., Blanden, R. V., and Zinkernagel, R. M. *Transplant. Rev.* **29**, 89, 1976.
2. Zinkernagel, R. M., and Doherty, P. C. *Nature* **251**, 547, 1974.
3. Koszinowski, U., and Ertl, H. *Nature* **255**, 552, 1975.
4. Gardner, T., Bowern, N. A., and Blanden, R. V. *Eur. J. Immunol.* **4**, 68, 1974.
5. Blank, K. J., Freedman, H. A., and Lilly, F. *Nature* **260**, 250, 1976.
6. Zweerink, H. J., Courtneidge, S. A., Skehel, J. J., and Crumpton, M. J. *Nature* **267**, 354, 1977.
7. Doherty, P. C., Effros, R. B., and Bennink, J. *Proc. Natl. Acad. Sci. USA.* **74**, 1209, 1977.
8. Blank, K. J., and Lilly, F. *Nature* **269**, 808, 1977.
9. Gomard, E., Duprez, V., Reme, T., Colombani, M. J., and Levy, J. P. *J. Exp. Med.* **146**, 909, 1977.
10. Goulmy, E., Termijtelen, A., Bradley, B. A., and van Rood, J. J. *Nature* **266**, 544, 1977.
11. Dickmeiss, E. Soeberg, B., and Svejgaard, A. *Nature* **270**, 526, 1977.
12. McMichael, A. J., Ting, A., Zweerink, H. J., and Askonas, B. A. *Nature* **270**, 524, 1977.
13. Ewan, P., and Lachmann, P. *Clin. Exp. Immunol.* **30**, 22, 1977.
14. Perrin, L. H., Zinkernagel, R. M., and Oldstone, M. B. A. *J. Exp. Med.* **146**, 949, 1977.
15. Zinkernagel, R. M. *J. Exp. Med.* **143**, 437, 1976.
16. Schmidt-Verhulst, A., and Shearer, G. M. *J. Exp. Med.* **142**, 914, 1975.

SELF-RESTRICTED CYTOTOXICITY AGAINST ACUTE MYELOID LEUKEMIA CELLS

R. T. D. Oliver and S. K. Lee

Recently, Zarling *et al.* (1) reported one patient with acute myeloid leukemia (AML) whose remission lymphocytes developed significant cytotoxicity against autologous leukemic blast cells. Development of cytotoxicity required *in vitro* sensitization with a mixture of mitomycin-C-inactivated autologous leukemic blasts and allogeneic normal lymphocytes. We have adapted this assay to study patients who have been undergoing immunotherapy with BCG or BCG plus irradiated allogeneic blast cells (2, 3).

MATERIALS AND METHODS

Clinical details including diagnostic criteria and methods of treatment have been reported (2-4).

Details of the cytotoxic assays are given elsewhere (5). In brief, 5×10^6 remission lymphocytes prepared by Ficoll-Hypaque separation from patients with AML were mixed with 2.5×10^6 autologous blast cells (inactivated by treatment with 50 μg mitomycin C at $37°$C for 30 min, washed twice) and 2.5×10^6 allogeneic normal lymphocytes (inactivated by treatment with 25 μg mitomycin C per 5×10^6 cells in 1 ml at $37°$C for 30 min, and washed twice). After six to seven days in culture, using RPMI 1640 medium and 10% AB serum supplemented with glutamine, in 5% CO_2 and air at $37°$C, the cells were harvested and

TABLE 1. CML Experiments with Remission Lymphocytes

Time in Remission (months)	No.	% ^{51}Cr Release from autologous leukemic blast cells				% ^{51}Cr Release from autologous lymph lymphocytes[a] (Priming technique[d])				% ^{51}Cr Release from allogeneic lymphocytes[a]			
		1	2	3	4	1	2	3	4	1	2	3	4
>6 (8-36)	9	3.2[b] (0-8)	6.3 (0-30)	27 (9-60)	5.6	0	1 (0-3)	1 (0-6)	1 (0-3)	2.5 (0-11)	35 (10-60)	27 (6-74)	ND[c]
<6 (2-5)	5	1.4 (0-5)	1.1 (0-6)	5.2 (0-19)	1.2	0	0	0	0	1 (0-10)	10 (3-19)	9.6 (3-26)	ND[c]

[a] Cultured with PHA for three days prior to labeling.
[b] Mean % ^{51}Cr release (figures in brackets indicate range).
[c] ND = Not done.
[d] Culture of remission lymphocytes for seven days with (1) autologous leukemic blasts; (2) allogeneic lymphocytes; (3) autologous blasts and allogeneic lymphocytes; (4) allogeneic blast cells.

5×10^5 lymphocytes were incubated with 1×10^4 ^{51}Cr-labeled autologous blast cells in a total volume of 0.2 ml of RPMI 1640 medium and 10% AB serum, in V-shaped Cooke microtiter plates (M220). After four hours, the plates were spun and 0.1 ml supernatant assayed for ^{51}Cr release. Specific release was calculated by the standard formula:

$$\frac{\text{Test release} - \text{spontaneous release}}{\text{Total release} - \text{spontaneous release}} \times 100$$

RESULTS

A summary of reactivities against three targets of primed remission lymphocytes from 14 patients is shown in Table 1. The patients were divided into two groups on the basis of duration of complete remission at the time of testing. It is apparent that overall cytotoxicity, whatever the target or priming method, is higher in those patients in remission for longer than six months. Cytotoxicity against autologous leukemic blast cells was greatest after priming with a mixture of autologous blasts plus allogeneic lymphocytes (column 3), though this procedure produced negligible cytotoxicity against autologous remission lymphocytes, and less cytotoxicity against allogeneic lymphocytes than was achieved by priming with allogeneic lymphocytes alone (column 2). It is important to note that priming with allogeneic AML blasts (column 4) produced no more cytotoxicity against autologous blasts than was achieved by priming with allogeneic lymphocytes. This was substantially less than achieved by priming with the mixture of autologous blasts plus allogeneic lymphocytes (column 3).

Nature of Effector Cells

Five experiments were performed with unseparated, T-enriched, and B-enriched primed cells produced by two methods of separation. No reactivity against autologous blast cells was demonstrable in B-enriched populations, while reactivity almost equal to that of unseparated effectors could be demonstrated in T-enriched populations (Table 2).

Nature of Helper Antigen on Third Party Allogeneic Cells

Though it was not possible to select a variety of third-party helper cells matched for different regions of the HLA complex with responding remission

TABLE 2. CML Experiments with Fractionated Effector Cells and Autologous Leukemic Blast Targets

	% ^{51}Cr Release (E:T=50:1) Expt. No				
Effector cells	1	2	3	4	5
Unseparated	45	51	14	28	4
T-enriched	44[a]	28[a]	12[a]	27[b]	10[b]
B-enriched	0[c]	0[c]	0[c]	0[d]	0[d]

[a] Effector population treated with anti-CLL serum (absorbed with platelets and thymocytes) plus complement.
[b] Sheep erythrocyte rosetting positive effector cells.
[c] Effector population treated with anti-T-lymphocyte serum (absorbed with CLL cells) plus complement.
[d] Sheep erythrocyte rosetting negative effector cells.

lymphocytes, preliminary evidence from lymphoblastoid cell lines suggests that genes associated with the HLA-D region are more important than those of the HLA-A and B loci. Daudi, a lymphoblastoid cell line that lacks HLA-A, B, and C locus antigens, but carries HLA-D determinants, was able to act as a third-party helper cell. On the other hand, the Molt-4 line carrying HLA-A and B antigens, but not HLA-D antigen, was unable to help generate cytotoxic cells active against autologous blast cells (Table 3). Of interest was the finding that priming with Daudi, with or without autologous blast cells, generated significant cytoxicity against Daudi. This indicates that cell-mediated cytotoxicity reactions are possible without matching effector and target cells for HLA-A, B, or C locus antigens.

TABLE 3. Lymphoblastoid Cell Lines as Helper Cells in Autologous Blast Cell Cytotoxicity Assay

		Targets		
Priming technique	No. of assays	PHA-Transformed autologous lymphocytes (%)[a]	Autologous leukemic blasts (%)	Daudi (%)
Autologous blast cells	34	0.1	3.7	10
Daudi	16	0.3	2.5	52
Daudi plus autologous blasts	16	0.2	26	41
Molt-4 plus autologous blasts	3	0.3	2	ND

[a] Figures indicate average specific ^{51}Cr release after background subtraction.

Nature of Target Antigen

Three lines of evidence suggest that the target antigen is not present on allogeneic leukemic myeloblasts.

(i) The priming of remission lymphocytes with allogeneic leukemic blast cells elicits marginally less cytotoxicity than allogeneic lymphocytes alone (Table 1, columns 3 and 4) and substantially less than the priming with autologous blasts plus allogeneic lymphocytes (Table 1, column 3).

(ii) Employing pairs of patients and performing cross-over experiments using lymphocytes and blasts of one patient as helper cells for the second patient, and *vice versa*, failed to provide evidence that allogeneic blasts are able to prime for cytotoxicity against specific autologous blasts (Table 4).

(iii) Cold target inhibition studies (Table 5) show that it is possible to inhibit competitively specific cytotoxicity with autologous blast cells only. None of the five allogeneic blast cells tested caused any reduction in cytotoxicity even though one was matched for three of four HLA-A and B antigens present on the specific target cell.

DISCUSSION

The idea of immunological stimulation by one determinant augmenting responses to an otherwise nonimmunogenic determinant is not new. In experimental animals, the use of carrier molecules to stimulate antibody response to haptens is standard procedure and the interdependence of T and B cells in such responses is well documented (for review, see 5). In man, Eijsvoogel *et al.* (6) were the first to demonstrate the interaction between two determinants necessary for the generation of cytotoxic T cells against transplantation antigens. These authors also showed that it was not necessary for the HLA lymphocyte-activating determinant to be on the same cell as the HLA-A, B, or C target antigen (7).

The data presented here confirm the observations of Zarling *et al.* (1) and Sondell *et al.* (8) who were the first to apply the three-cell principle of Eijsvoogel to patients with leukemia. However, the present data provide the first evidence that this system is also dependent on interaction between antigens of the HLA region, like Eijsvoogel (6, 7) reported for allogeneic CML. It was demonstrated convincingly that antileukemic cytotoxicity is mediated by T lymphocytes. Though this result is hardly surprising, only the study of Jondal *et al.* (9) of a single patient with Burkitt's lymphoma has provided convincing evidence for tumor-specific T-cell cytotoxicity in man.

Immunotherapy with allogeneic-irradiated leukemic blasts has been under investigation in man for more than 12 yr (for review, see 3). The evidence pre-

TABLE 4. Search for Target Antigens in Autologous Blast Cell Cytotoxicity[a]

	Priming Technique				
Lymphocyte donors	Autologous blasts	Allogeneic lymphocytes	Allogeneic lymphocytes plus autologous blasts	Allogeneic blasts	Allogeneic blasts plus allogeneic lymphocytes
T.W.	0[b]	0	9	0	0
L.F.	0	0	16	0	0
M.B.	ND	0	23	7	5
T.W.	ND	1	51	0	0

[a] Autologous blasts are the target cells.
[b] Percent specific lysis.

TABLE 5. Cold Target Cell Inhibition Experiments

	Target cell mixtures			
Expt. No.	Autologous blasts-^{51}Cr (1 x 10^4)	Autologous blasts-^{51}Cr (1 x 10^4) plus autologous blasts (1 x 10^4)	Autologous blasts-^{51}Cr (1 x 10^4) plus autologous remission lymphocytes (1 x 10^4)	Autologous blasts-^{51}Cr (1 x 10^4) plus allogeneic blasts (1 x 10^4)
1	34	5	36	31, 34, 37
2	21	1	18	(23)[d], 19
3	12	0	6	12, 18
4	43	30[a]	41[a]	35[a,e]
		6[b]	39[b]	36[b,c]
		0[c]	40[c]	33[c,e]

[a] Cold targets 1 x 10^3. Percent specific lysis.
[b] Cold targets 1 x 10^4.
[c] Cold targets 4 x 10^4.
[d] Cold targets = Daudi.
[e] Cold targets = allogeneic PHA blasts.

sented in Tables 1, 4, and 5 suggests that there is a unique leukemia-specific antigen restricted to the patients' own blast cells. This may be the reason for the failure of immunotherapy with allogeneic tumor cells (3).

It is premature to speculate whether there is HLA restriction influencing recognition of the target antigen, as demonstrated for most mouse cytotoxicity reactions (10), or derepression of genes that are not normally expressed, as found for some tumor antigens in the mouse (11). More extensive crosstesting of pairs of patients is essential for sorting out these possibilities.

CONCLUDING STATEMENT

In a three-cell system, remission lymphocytes of AML patients, inactivated autologous leukemic blast cells, and allogeneic lymphocytes were set up in short-term cultures. Effector T lymphocytes cytotoxic for the specific autologous blast cells were generated in cultures derived from 11 of 14 patients studied. Experiments with Daudi and Molt-4 lymphoblastoid cell lines as third-party helper cells indicated that HLA-D locus incompatibility was required for effective helper function in this system. Cold target inhibition experiments, cross-reaction studies between pairs of patients, and experiments with allogeneic leukemic blast cells as priming stimulus, suggest that the target antigen is only present on autologous leukemic blasts.

References

1. Zarling, M. J., Raich, P. C., McKeough, M., and Bach, F. H. *Nature* **262**, 691, 1976.
2. Powles, R. L., Russell, J., Lister, T. A., Oliver, R. T. D., Whitehouse, J. M. A., Peto, R., Chapuis, B., Crowther, D., and Alexander, P. *Brit. J. Cancer* **35**, 265, 1977.
3. Oliver, R. T. D. *Progr. Immunol.* **3**, 572, 1977.
4. Crowther, D., Powles, R. L., and Bateman, C. J. T. *Brit. Med. J.* **1**, 131, 1973.
5. Hogg, N., Koszinowski, U. and Mitchison, N. A. *Progr. Immunol.* **3**, 532, 1977.
6. Eijsvoogel, V. P., du Bois, M. J. G. S., Melief, C. J. M., Zeylemaker, W. P., Raat-Konig, L. and Groot-Kooy, L. *Transplant. Proc.* **5**, 1301, 1973.
7. Eijsvoogel, V. P., de Bois, M. J. G. S., Meinesz, A., Bierhorsch-Eijlander, W. P., Zeylemaker, W. P. and Schellekens, T. Th. A. *Transplant. Proc.* **5**, 1675, 1973.
8. Sondell, P. M., O'Brien, C., Porter, L., Schlossman, S. F., and Chess, L. *J. Immunol.* **117**, 2197, 1976.
9. Jondal, M., Svedmyr, E., Klein, E., and Singh, S. *Nature* **255**, 405, 1975.
10. Zinkernagel, R. M., and Doherty, P. C. *J. Exp. Med.* **141**, 1426, 1975.
11. Garrido, F., Festenstein, H. and Shirrmacher, V. *Nature*, **261**, 705, 1976.

REGULATORY ROLE OF MACROPHAGES IN NORMAL AND NEOPLASTIC HEMOPOIESIS

Malcolm A. S. Moore and Jeffrey I. Kurland

The development of a semi-solid agar culture system for cloning murine granulocyte-macrophage committed stem cells (CFU-c) (1), and its subsequent adaptation for human studies (2), has resulted in the accumulation of much information on this committed stem cell population in health and disease.

Granulopoiesis *in vitro* is dependent on diffusible activities termed colony-stimulating factors (CSF). Addition of CSF to cultures at concentrations as low as $10^{-12} - 10^{-14}$ M will promote colony formation. The factor is not a simple inducing agent, since its presence is continually required throughout the growth of the colony in addition to promoting CFU-c survival, proliferation, and differentiation (3). Recently, purified CSF preparations have also been shown to act as macrophage growth stimulating or mitogenic factors, stimulating extensive peritoneal macrophage or blood monocyte proliferation in agar culture (4, 5). Extensive functional and biochemical studies on CSFs obtained from a number of sources have revealed considerable heterogeneity of material with colony-stimulating activity.

Recognition that CSF is produced by monocytes and macrophages (6-8), and that it acts to promote increased monocyte production and macrophage proliferation, introduces the problem of mechanisms designed to counterbalance this positive feedback drive. A number of mechanisms have been revealed in *in vitro* studies and many, if not all, may be of physiological significance *in vivo*. The functional heterogeneity of the phagocytic mononuclear cell population must first be considered, since marked variation in CSF-producing capacity exists. "Virgin" macrophages developing in agar culture from CFU-c and macrophages generated in continuous marrow culture are not constitutive producers of CSF; however, exposure of these cells to macrophage activating agents such as lipopolysaccharide or BCG rapidly induces CSF synthesis and secretion (Moore, M. A. S., unpublished observation). In this sense, CSF recruitment of additional monocytes and macrophages would not, ipso facto, lead to increased CSF production in the absence of an exogenous source of stimulation such as endotoxemia due to gram-negative bacterial infection. Neoplastic monocyte or macrophage cell lines also retain the capacity to produce CSF; however, in some cases the leukemic cell lines are constitutive producers and in other cases CSF production is observed only after LPS stimulation, suggesting retention of a degree of normal responsiveness by the transformed cells (9).

The regulatory interactions involving diffusible stimulatory and inhibitory activities elaborated by granulocytes, lymphocytes, and phagocytic mononuclear cells can clearly involve specific macromolecules or, alternatively, nonspecific modulating activities. Pharmacological studies have shown that prostaglandins of the E series (PGE) and other agents capable of elevating intracellular levels of c-AMP profoundly inhibit granulopoiesis and macrophage proliferation *in vitro* (10). Just as CSF promotes continued replication of the CFU-c and its progeny, PGE limits this effect by an opposing action on the responsiveness of the myeloid stem cell and its proliferative progeny to stimulation by CSF. Kurland and Moore (10, 11) have shown that prostaglandin synthesized by phagocytic mononuclear cells may be of central importance in the modulation of hemopoiesis. Measurement of prostaglandin E production by murine macrophages and human monocytes has been performed using a sensitive radioimmunoassay and has shown a linear relationship between the number of phagocytic mononuclear cells and the concentration of PGE in the conditioned medium (12). This observation explains the lack of correlation between the numbers of monocytes and macrophages used to stimulate granulocyte-macrophage colony formation and the incidence of colonies. Titration of varying numbers of adherent macrophages or blood monocytes as a source of stimulus for human or murine marrow CFU-c has clearly shown that colony formation is stimulated by low numbers of phagocytic mononuclear cells (0.05-2×10^5) and inhibited if higher concentrations are used. Parallel studies using monocytes or macrophages treated with indomethacin, a potent inhibitor of prostaglandin synthesis, have revealed a linear relationship between the number of colonies stimulated and the number of phagocytic mononuclear cells used as the source of CSF (12). These observations point to the unique ability of the macrophage to control the proliferation of its own progenitor cell by elaboration of opposing regulatory influences.

Utilizing three cellular systems, we have established that the synthesis and release of PGE by macrophages is determined by an afferent feedback mechanism involving macrophage surveillance of local CSF concentrations. In experiments using macrophages from strains of mice differing only in their responsiveness to endotoxin (LPS), it was found that LPS stimulated both CSF and PGE production by macrophages from normal C3HeB/FeJ, but not LPS-nonresponsive C3H/HeJ mice. However, both CSF and PGE synthesis by macrophages from C3H/HeJ mice was stimulated by Zymosan or Con A, and methylmannoside selectively prevented the inductive effect of Con A. In all cases, both constitutive and stimulated production of CSF preceded the active synthesis of PGE, and the addition of a soluble source of CSF alone stimulated PGE synthesis and circumvented the LPS-nonresponsiveness of C3H/HeJ macrophages. A number of continuous macrophage cell lines constitutively synthesized PGE and CSF. Neoplastic macrophage cell lines that were not constitutive producers of either CSF or PGE could be induced to active synthesis following treatment with Zymosan, LPS, or PPD. In all cases, CSF production was followed by the rapid induction of PGE synthesis, and in the absence of CSF-promoting agents, PGE was induced simply by a soluble source of CSF.

We also investigated the homeostatic potential of the monocyte-macrophage effector role in hematopoiesis in terms of (i) the ability to increase its elaboration of a diffusible inhibitor of CFU-c proliferation in response to increasing concentrations of CSF, and (ii) whether this inhibition correlated with an increase in the elaboration of PGE by the mononuclear phagocytes. The species specificity of CSF action, whereby CSF of murine cell origin does not stimulate colony formation by human CFU-c (3), provided the basis of these investigations in which we utilized normal human bone marrow cells incorporated in soft agar cultures as the source of target-committed stem cells. Such cultures were stimulated by the presence of human CSF elaborated by 1×10^6 normal human peripheral blood leukocytes suspended in a 0.5% agar underlayer. Cell-free supernatants from 48-hr liquid cultures of 1×10^6 murine peritoneal macrophages per milliliter were prepared in the absence and presence of a potent source of murine CSF, provided by conditioned medium from a murine myelomonocytic leukemic cell line (WEHI-3). Neither the individual presence of the murine macrophage supernatant, nor the murine CSF to which the macrophages were exposed, had any effect on the human CSF-stimulated CFU-c proliferation when tested at 10% by volume in soft agar cultures. However, the peritoneal macrophage supernatant prepared in the continued presence of the murine CSF profoundly inhibited the proliferation of human CFU-c stimulated by the human leukocyte-derived CSF. the magnitude of human CFU-c inhibition was proportional to the concentration of murine CSF present during the active generation of the murine macrophage supernatant. After simple dialysis, these same macrophage supernatants lost all detectable inhibitory activity against human bone marrow CFU-c. These findings indicate that murine macrophages, which have otherwise no effect on exogenously stimulated human CFU-c, are induced by increasing concentrations of murine CSF to elaborate a low molecular weight, nonspecies specific inhibitor of committed granulocyte-macrophage stem cell proliferation.

The same concentrations of murine CSF that promoted the elaboration of CFU-c inhibitory factor, induced the coincident production of PGE by 1×10^5 murine peritoneal macrophages incubated under conditions identical to the preparation of the macrophage supernatants used in the previous experiment. The present findings are based on techniques available for assaying hematopoietic cell cloning *in vitro*, but are sufficient to suggest physiological relevance to the situation *in vivo*. In this regard, the mononuclear phagocyte is implicated as the central cellular element in the humoral control of granulopoiesis and monocyte-macrophage production. The ability of this cell population to elaborate both CSF and PGE, agents that exert an opposing proliferative influence on the committed granulocyte-macrophage stem cell (CFU-c), indicates that this regulation is based on a balance of both positive and negative feedback mechanisms. This model explains some of the opposing effects of the mononuclear phagocyte on hematopoietic cell proliferation.

Under basal conditions, appropriate levels of CSF elaborated by the monocyte-macrophage population (6-8) stimulate the committed stem cell to commence proliferation. The constitutive contribution of CSF to granulopoiesis and monocyte-macrophage production is rapidly increased under physiologically perturbed circumstances. Thus, after trauma or during acute viral or bacterial infections, increases in both systemic and local CSF concentrations rapidly ensue (13, 14). This increased CSF stimulates a greater number of stem cells to proliferate which, in addition to recruiting granulocytes, results in an expansion of the population of CSF-producing mononuclear phagocytes. Such a positive feedback control of myelopoiesis is ultimately restricted by the synthesis and release of PGE by the mononuclear phagocyte components of the myeloid clone. Just as CSF promotes continued replication of the CFU-c and its progeny, PGE limits this effect by an opposing action on the responsiveness of the myeloid stem cell and proliferating progeny to stimulation by CSF. Thus, the net proliferative potential of the myelopoietic clone is determined by the relative levels of CSF and PGE which, in turn, is controlled by a balance between both positive and negative feedback mechanisms of the mononuclear phagocyte.

Since two opposing feedback mechanisms have their origins in a common regulatory cell, the two mutually antagonistic feedback events are probably causally associated. Such a proposition is substantiated by the observations that progressive increases in CSF beyond a critical concentration are ultimately sensed by the macrophage and serve to stimulate the coincident production and release of PGE. The concentrations of macrophage-derived PGE increase in parallel with the local CSF concentration, thereby implicating the mononuclear phagocyte as a surveillance cell that functions to maintain myelopoiesis within appropriate limits. The extreme lability of the prostaglandin molecule (15) may provide a further physiological control, for which the continued presence of an elevated CSF stimulus may be a requisite for the maintenance of critical PGE concentration.

The recent advent of an *in vitro* cloning technique that permits the clonal proliferation of a population of murine-B lymphocytes (16, 17) provided a means of investigating the interactions between B lymphocytes and macrophages. By preventing intimate cell-cell contact, the semisolid nature of the culture system allowed an analysis of immune regulatory molecules elaborated by a particular accessory cell population under conditions of extremely low cell densities (18). B-lymphocyte proliferation is dependent upon the presence of 2-mercaptoethanol (19) and serum, as well as mitogens native to laboratory grade agar (20). As previously described, this population of B lymphocytes are heterogeneous with respect to size and density (16, 21), and the majority bear Fc receptors, IgM, and Ia antigens on their surface (17, 22). The effects of macrophages on B-lymphocyte clonal proliferation was investigated in a two-layer soft agar culture system where macrophage-B-lymphocyte interactions are prevented by the gel matrix. Adherent macrophage underlayers potentiated both the number and size of B-lymphocyte colonies in mouse spleen or lymph node cultures. Macrophage-depleted lymph node cells or neonatal spleen cells formed virtually no colonies on macrophage-free underlayers, but colony formation could be restored by an optimal number of macrophages. When greater than optimal numbers of macrophages were present, B-lymphocyte colony formation was inhibitied. The presence of indomethacin, a prostaglandin synthesis inhibitor, in the culture system markedly enhanced the number of B-lymphocyte colonies that developed in the presence of high numbers of macrophages. These results suggest that macrophage-derived PGE suppresses B-lymphocyte proliferation and serves to counteract the stimulatory factor produced by macrophages.

The factor that stimulates B-cell clonal proliferation was detected in culture supernatants of adherent cells of both thioglycollate-induced peritoneal exudates and noninduced peritoneal washings, but not in medium conditioned by the nonadherent components of the mouse peritoneal cavity. Similarly, significant stimulatory activity was evident in culture fluids of adherent but not nonadherent spleen cells, lung, and, to a lesser extent, liver and bone marrow. The cells of origin of the B-lymphocyte stimulatory factor in peritoneal exudates and washings were identified as macrophages on the basis of their characteristic morphology and histochemical staining for cytoplasmic nonspecific esterase. Conceivably, the mononuclear phagocyte of spleen, alveolar macrophage, Kupffer cells, and bone marrow histiocytes are the sources of this factor in the other tissues. Furthermore, though it has been reported that adherent fibroblast cell lines can in part share the macrophage's ability to augment antibody responses (23), attempts to detect an activity that stimulated B-cell clonal proliferation in soft agar by culture fluids from a number of fibrosarcoma and monocytic-histiocytic cell lines have been so far unsuccessful. In this regard, conditioned medium from the murine myelomonocytic cell line (WEHI-3), which elaborates both colony-stimulating factor (24) and lymphocyte-activating factor (25), cannot substitute for the macrophage in the initiation of B-cell colony formation. These findings suggest the nonidentity of this macrophage-derived B-cell clonal stimulating activity with other known factors that alter hemopoietic and lymphoid function.

The implications of this dualistic role of macrophages in regulation of B-cell proliferation is illustrated by the influence of LPS on B-lymphocyte clonal proliferation. LPS is a direct B-cell mitogen and hence can markedly enhance B-lymphocyte colony formation in the presence of low numbers of macrophages. With increasing numbers of macrophages, LPS addition to cultures leads to reduction in B-lymphocyte cloning, which is reversed by treatment of macrophages with indomethacin. Since LPS leads directly to macrophage activation with induction of CSF synthesis and subsequent CSF-dependent macrophage prostaglandin synthesis, the proliferative response of B cells can be influenced in either a positive or negative way by LPS treatment, depending upon cell density and ratio of B cells to macrophages.

Considerable evidence is now available that macrophages represent an important component in both immunologically specific and nonspecific resistance against tumor cells (26-30). Though this may appear to be dependent upon a population of activated macrophages that is strikingly effective in killing tumor cells *in vitro* (26-28), a number of reports document that nonactivated macrophages found in sterile inflammatory exudates are incapable of tumor cell cytotoxicity but nonetheless exert a profound inhibitory effect on the proliferation of tumor cells *in vitro* (29-31). In all previous studies of macrophage-mediated cytostatis of tumor cells, the endpoint determination has been the suppression of radioactive-labeled DNA precursor incorporation. However, the findings of Opitz *et al.* (32, 33) that macrophages release endogenous thymidine indicate that the decrease in (^{3}H) thymidine or (^{125}I) iododeoxyuridine incorporation may not truly reflect an inhibition of tumor-cell proliferation and may only be due to altering the size of the thymidine pools. Similar findings have recently been reported by Unanue *et al.* (34). Therefore, a multiparameter analysis is essential to understanding how macrophages can effect tumor-cell proliferation.

We elected to investigate this phenomenon of nonimmune macrophage-mediated cytostasis using target cell populations of leukemic hemopoietic cells proliferating both in conventional liquid cultures, as well as in semi-solid culture; the latter system was designed to quantitate the clonal proliferation of leukemic cells and the ability of diffusible factor(s) from underlayers of adherent peritoneal macrophages to mediate cytostasis. In addition to (^{3}H) TdR incorporation, colony-forming ability, and viable cell numbers as indices of cell proliferation, cell cycle analysis by flow cytofluorometry was used to establish the existence and nature of hemopoietic cell cycle inhibition.

Nonactivated, nonimmune mouse peritoneal macrophages exerted a profound inhibition on neoplastic hemopoietic cell proliferation *in vitro*. This effect was dependent upon the number of adherent macrophages present in cultures of hemopoietic tumor cells The two-layer soft agar culture system allows low numbers of leukemic cells to be suspended in a semisolid matrix and their clonal proliferation to be quantitated. The ability of underlayers of macrophages to inhibit leukemic cell colony formation indicates that physical contact is not required and that such an effect can be mediated by diffusible factor(s) produced by macrophages. Similarly, inhibition of (^{3}H) thymidine incorporation, cell numbers, and mitotic indices, each manifested with a coincident retention of cell viability, can be documented in conventional liquid cultures in the presence of macrophages. Under these conditions, analysis by acridine orange straining and flow cytofluorometry indicated that the tumor cells were blocked primarily in the G_1 phase of the cell cycle. These findings are similar to the report of Keller *et al.* (35) who, by using impulse microfluorometry, observed a reduction in the ethidium bromide intercalated into the DNA of the isolated nuclei of rat tumor cells inhibited by macrophages, which suggested a G_1 cell-cycle block. Thus, evidence is presented that the mode of action of nonactivated macrophages on hemopoietic tumor cell proliferation is to retard the tumor cell in a particular phase of the cell cycle. Furthermore, the true cytostatic nature of this macrophage effect is indicated by the reversibility of suppressed cell growth and mitotic index and by the observation that tumor cells begin to center a normal cycle distribution shortly after being removed from the macrophage influences. Such cytostatic reversibility is dependent upon the initial ratio of macrophages to leukemic cells, the length of time in which the leukemic cells have been in the presence of macrophages, and the sensitivity of the particular tumor cell to inhibition by macrophages.

Of particular interest was the correlation between the sensitivity of the hemopoietic tumor cells to inhibition by macrophages and the position in the cell cycle in which they were inhibited. Three human hemopoietic cell lines (CCRF-CEM, WIL-2, and K562) moderately sensitive to macrophages, as well as the murine leukemic cell line (WEHI-3), which was the most sensitive to inhibition by macrophages, exhibited a G_1 cell-cycle block, the magnitude of which was dependent upon the initial ratio of macrophages to target cells. In contrast, a murine T-cell lymphoma EL-4, the least sensitive to inhibition by macrophages, exhibited a block at equivalent macrophage numbers. However, at higher macrophage: leukemic cell ratios at which CCRF-CEM, WIL-2, K562, and WEHI-3 exhibited a complete G_1 block, EL-4 began to shift its G_2-restricted cell cycle position to that of a G_1 block. This suggests that EL-4 can be blocked in one of two places in the cell cycle, the position of which depends on the number of macrophages present.

This cytostatic effect of macrophages on neoplastic cell proliferation does not require cell contact and is mediated by a diffusible, dialyzable activity whose production is indomethacin-sensitive. Correlation of prostaglandin E production as detected by radioimmunoassay with the inhibitory activity of macrophage-conditioned medium on tumor proliferation indicate that indeed prostaglandin is the activity responsible for macrophage-mediated tumor cell cytostasis. These results suggest that neoplastic hemopoietic cells retain some degree of responsiveness to macrophage regulation. Indeed, leukemic cells appear even more sensitive to inhibition by macrophages than their nontransformed counterparts (12), suggesting that biosynthesis of prostaglandin E by macrophages may play a role in host defense against neoplasia.

Acknowledgment

This work was supported by grant number CA-19052 from the National Institutes of Health.

References

1. Bradley, T. R., and Metcalf, D. *Aust. J. Exp. Biol. Med. Sci.* **44**, 287, 1966.
2. Pike, B. L., and Robinson, W. A. *J. Cell. Physiol.* **76**, 77, 1970.

3. Metcalf, D., and Moore, M. A. S. "Haemopoietic Cells." ASP-Biological and Medical Press (North-Holland Division), Amsterdam, 1971.
4. Stanley, E. R., Cifone, M., Heard, P. M., and Defendi, V. *J. Exp. Med.* **143**, 631, 1976.
5. Lin, H. S., and Freeman, P. G. *J. Cell. Physiol.* **90**, 407, 1977.
6. Moore, M. A. S., and Williams, N. *J. Cell. Physiol.* **80**, 195, 1972.
7. Golde, D. W., and Cline, M. J. *J. Clin. Invest.* **51**, 2981, 1972.
8. Moore, M. A. S., Spitzer, G., Metcalf, D., and Penington, D. G. *Brit. J. Haematol.* **27**, 47, 1974.
9. Ralph, P., Broxmeyer, H. E., and Nakoinz, I. *J. Exp. Med.* **146**, 611, 1977.
10. Kurland, J., and Moore, M. A. S. *Exp. Hematol.* **5**, 357, 1977.
11. Kurland, J., and Moore, M. A. S. *In* "Experimental Hematology Today" (S. Baum and G. Ledney, Eds.), p. 51, Springer-Verlag, Berlin, 1977.
12. Kurland, J., Bockman, R., Broxmeyer, H. E., and Moore, M. A. S. *Science* **199**, 552, 1978.
13. McNeil, T. A. *Immunology* **18**, 61, 1970.
14. Eaves, A. C., and Bruce, W. R. *Cell Tissue Kinet.* **7**, 19, 1974.
15. Anderson, N. H., and Ranwell, P. W. *Arch. Int. Med.* **133**, 30, 1974.
16. Metcalf, D., Nossal, G. J. V., Warner, N. L., Miller, J. F. A. P., Mandel, T. E., Layton, J. E., and Gutman, G. A. *J. Exp. Med.* **142**, 1534, 1975.
17. Metcalf, D., Warner, N. L., Nossal, G. J. V., Miller, J. F. A. P., Shortman, K., and Rabellino, E. *Nature* **255**, 630, 1975.
18. Kurland, J., Kincade, P. W., and Moore, M. A. S. *J. Exp. Med.* **146**, 1420, 1977.
19. Metcalf, D. *J. Immunol.* **116**, 635, 1977.
20. Kincade, P. W., Ralph, P., and Moore, M. A. S. *J. Exp. Med.* **143**, 1265, 1977.
21. Metcalf, D., Wilson, J. W., Shortman, K., Miller, J. F. A. P., and Stocker, J. *J. Cell Physiol.* **88**, 197, 1976.
22. Kincade, P. W., and Ralph, P. *Cold Spring Harbor Symp.* **41**, 245, 1977.
23. Moller, G., Lemke, H., and Opitz, H. G. *Scand. J. Immunol.* **5**, 269, 1976.
24. Ralph, P., Moore, M. A. S., and Nilson, K. *J. Exp. Med.* **143**, 1528, 1976.
25. Hacker, M. P., Lachman, L., Blyden, G. T., and Handschumacher, J. *Fed. Proc.* **36**, 1300, 1977.
26. Alexander, P., and Evans, R. *Nature New Biol.* **232**, 76, 1971.
27. Hibbs, J. B. *J. Natl. Cancer Inst.* **53**, 1487, 1974.
28. Holtermann, O. A., Klein, E., and Casale, G. P. *Cell. Immunol.* **9**, 339, 1975.
29. Calderon, J., Williams, R. T., and Unanue, E. R. *Proc. Natl. Acad. Sci. USA* **71**, 4273, 1974.
30. Keller, R. *Brit. J. Cancer* **30**, 401, 1974.
31. Kirchner, H., Holden, H. T., and Herberman, R. B. *J. Natl. Cancer Inst.* **55**, 971, 1975.
32. Opitz, H. G., Niethammer, D., and Flad, H. D. *Cell. Immunol.* **16**, 379, 1975.
33. Opitz, H. G., Niethammer, D., and Jackson, R. C. *Cell. Immunol.* **18**, 70, 1975.
34. Unanue, E. R., Beller, D. I., Calderon, J., Kiely, J. M., and Stadecker, M. J. *Am. J. Pathol.* **85**, 465, 1976.
35. Keller, R., Bregnard, A., Gehring, W. J., and Schroeder, H. E. *Exp. Cell. Biol.* **44**, 108, 1976.

SERUM HEMATOPOIETIC INHIBITORS AND TUMOR NECROSIS FACTOR

M. A. S. Moore, R. Shah, and S. Green

The relationships between the anticancer defenses of the host and the function of both the reticuloendothelial and immune system have been repeatedly demonstrated over the past 25 yr. Alexander and Evans (1) reported that endotoxin did not affect the growth of lymphoma or sarcoma cells in tissue culture but did activate macrophages, rendering them cytotoxic to sarcoma cells. Parr *et al*. (2) found that endotoxin was without effect on tumor cell growth *in vivo* but that appreciable inhibition of tumor growth occurred if macrophages previously activated by exposure to endotoxin were injected along with the tumor cells. Hibbs (3) found that BCG-activated macrophages contain a substance capable of selectively killing tumor cells, and Suter (4) reported that pretreatment with BCG rendered mice hypersensitive to endotoxin. This reactivity was not related to delayed tuberculin hypersensitivity, since production of a similar state of hyperreactivity was brought about with cord factor, a waxy noninfectious glycolipid from virulent tubercle bacilli. Younger and Stinebring (5) succeeded in potentiating interferon production in mice by the sequential administration of BCG and endotoxin, each of which is an interferon inducer per se. Carswell *et al*. (6) found that serum from mice treated sequentially with BCG and endotoxin was capable of inducing necrosis in experimental tumors. Since serum from mice treated with either BCG or endotoxin alone contained interferon but had no antitumor activity, and since no correlation was found between the tumor necrosis including activity in the serum and the interferon titer, it was concluded that the antitumor activity was not due to interferon. Further, the antitumor activity of BCG-endotoxin serum was judged not to be due to the presence of residual endotoxin since much less endotoxin was needed for the induction of antitumor activity in the serum than for the production of necrosis in tumors after direct endotoxin injection. In addition, the antitumor response was evoked more rapidly when the BCG-endotoxin serum was used than with endotoxin itself.

Green *et al*. (7) partially purified this serum tumor necrosis factor (TNF) from pools of 100 ml of serum from BCG-endotoxin-treated mice. All fractions collected during the purification were bioassayed for TNF activity with a standard *in vivo* Meth A assay (6). The serum pool was diluted with an equal volume of sterile 0.15 M NaCl and solid $(NH_4)_2SO_4$ was added to the supernatant solution to raise the saturation up to 70%. The precipitate was dissolved, dialyzed against phosphate-buffered saline (PBS) until free of $(NH_4)_2SO_4$. This fraction was placed on a 2.5 x 100 cm column of sephadex G-100 gel. TNF activity was found in the first protein peak eluted. The protein under peak 1 was placed on a G-200 Sephadex gel column and three protein peaks were eluted. The TNF activity was found in the sample collected under the second peak (G-200 II).

Further purification of TNF was carried out on polyacrylamide gels. A Canalco apparatus was modified to carry 3-4 mg of protein per gel and aliquots of the G-200 II preparation were placed on the gels and after electrophoresis, one of the G-200 II containing gels was stained for protein with Canalco aniline blue-black dye. A second gel containing G-200 II was stained for glycoprotein using the periodate-Schiff reaction of Zacharius (8) and each remaining gel was cut into three 5-cm segments (A, anodic; B, center; C, cathodic) and the protein in each segment was extracted with saline. TNF activity was present in the center segment (B). This segment had four major protein bands and stained positively for carbohydrate, indicating the presence of glycoproteins. The proteins in the B segment were separated by slicing the gels into smaller segments (i.e., B_1, B_2, B_3, and B_4) and the protein was extracted from each TNF was found in the first (B_1) and second (B_2) disks. Upon reelectrophoresis, we found that the protein in B separated into four bands that were identical with the original four. The second (B_2) separated into three bands, the third (B_3) separated into two bands, the last (B_4) migrated as a single band. Aliquots of G-200 II, G-200 II-B_1, G-200 II-B_2, and G-200 II-B_3 were electrophoresed on cellulose acetate strips. All the protein in the G-200 II-B_1, B_2, and B_3 fractions migrated as α-globulins and demonstrated that each was separating on the basis of molecular weights when electrophoresed on polyacrylamide gels. The molecular weight of the G-200 II-B fraction was determined on a G-200 Sephadex column and was 150,000 daltons. A more precise molecular weight determination was made in the model E analytical ultracentrifuge. Two bands of sedimenting proteins were observed that had sedimentation values (S 20 W) between 6 and 8. These values correspond to an average molecular weight of about 150,000 (7).

Qualitative examination of the phyhsical properties of the B_1-B_2 fraction showed that: TNF activity is not lost after 1 hr at 56°C but was completely destroyed after 1 hr at 70°C; repeated freezing and thawing did not diminish biological activity. Biochemical analysis showed the following: enzymes—no measurable acid phosphatase, alkaline phosphatase, β-glucuronidase, α-galactosidase, NADase, neuraminidase, or lysozyme; carbohydrates—3.3% sialic acid, 3.0% galactosamine, 0.5% glucose, and 0.33% fucose; fatty acids—0.1% myristic, 0.53% palmitic, 0.1% oleic, and 0.13% stearic; nucleic acids—no measurable DNA or RNA.

TNF is active against a variety of cells in culture. It is cytotoxic for mouse L-cells (NCTC clone 929), cytostatic for mouse Meth A cells, and has no effect on mouse embryo fibroblasts (6). Recently, Helson *et al.* (9) tested the effect of partially purified TNF against three neuroblastoma lines, a fibroblast line derived from the bone marrow of a child with neuroblastoma, and a human melanoma cell line. Only the melanoma cells in culture were sensitive to TNF and this sensitivity took the form of complete cytostasis. When macrophages are nonspecifically "activated" by agents such as BCG or *C. parvum* (CP), endotoxin, and certain protozoans, they acquire selective toxicity for malignant cells. The fact that TNF showed discriminatory *in vivo* and *in vitro* toxicity for tumor cells suggests that TNF may be a mediator in the selective cytotoxicity of activated macrophages. In addition, the tissue damage seen in TNF-treated Meth A tumors was not preceded by injury to the vasculature, indicating that hemorrhage was not a prerequisite for necrosis.

The protocol for induction of TNF activity parallels very closely that used to elicit activities regulating proliferation and differentiation of granulocytes, monocytes, and macrophages.

Bone marrow progenitor cells (CFU-c) proliferate and differentiate to form colonies of macrophages and/or granulocytes *in vitro* when stimulated by colony-stimulating activity (CSA). The cells capable of elaborating CSA have been identified as belonging to the monocyte-macrophage series (10) and attempts to characterize this activity revealed that it is composed of a heterogeneous group of molecules (11). Extensive characterization of CSA from human urine was shown to be a sialic acid containing glycoprotein (12).

The presence of CSA is demonstrable in the serum and urine of normal humans and mice (13), and is elevated following viral or bacterial infections (14, 15). In addition, the elevation in serum level of CSA has also been reported following injections of bacterial endotoxin or antigen in experimental mice (16,17).

Fluctuations in CSA content of various tissues are also observed in mice following endotoxin treatment and the temporal relationship of the magnitude of the increase in tissue production to serum elevation of CSA is complex (18). In such tissues as lung and spleen, the measurable CSA content always equalled or exceeded prestimulatory levels; however, distinctly different forms of CSA were observed as a function of time after endotoxin treatment of tissue of origin. CSA extracted from postendotoxin mouse lung tissue or *in vitro* lung-conditioned medium appeared to be glycoproteins, for like human urinary CSA, they were inactivated by proteases and altered in charge, but not inactivated by sialidase (19).

It has been proposed that endotoxin may be of central importance in regulation of granulopoiesis and monocyte production due to its capacity to stimulate

CSA-release and synthesis by the widely disseminated phagocytic mononuclear cell population. The existence of negative feedback control of granulopoiesis has been postulated as a requirement to counterbalance the positive feedback mediated by CSA, and numerous reports attest to the existence of humoral granulopoietic inhibitors.

The capacity of serum to stimulate granulocytic colony formation *in vitro* is determined by CSA content and by at least two other serum factors. A potentiating factor is present in the serum of mice and other species which enhances CSA action (20). Conversely, at high concentrations the serum from many strains of mice can inhibit colony formation even in the presence of adequate CSA concentrations due to the presence of one or more inhibitory factors in such sera. One inhibitory activity in mouse serum has been shown to be partially removable by precipitation following dialysis against water (21, 22), heating (20), or fat solvent extraction (23). Serum inhibitory activity was lowered by heparin or by starvation (24) and, from these observations, it has been suggested that the inhibitors are probably lipoprotein in nature. Metcalf and Russell (25) reported that inhibitory activity of mouse serum for granulocytic colony formation was separable by flotation centrifugation into a very light density fraction (VLD). On gel filtration, the inhibitory material had an apparent molecular weight of 250,000 and, on electrophoresis, was localized as a single peak near the albumin region. These inhibitory preparations exhibited lack of specificity and inhibited colony formation by normal B lymphocytes as well as by various leukemic cell lines. Some indirect evidence supports the view that inhibitory lipoprotein may play some role in regulation of myelopoiesis. Serum inhibitor levels rapidly declined after whole body irradiation (26) and this decrease could be prevented by partially shielding hematopoietic tissue or by the injection of syngeneic marrow cells immediately after irradiation (27). The existence of more than one serum inhibitor is suggested by the observation that upon gel filtration on Sephadex G-150 two distinct regions of inhibitory activity were found (28). One activity was in the excluded fraction and appeared to be a lipoprotein, the second was in the included fraction and was not a lipoprotein. While the majority of activities capable of inhibiting hematopoiesis *in vitro* have yet to have demonstrable *in vivo* action, at least four agents with defined *in vivo* action are present in normal serum and can inhibit hematopoiesis *in vitro*: (a) glucocorticoids that inhibit colony formation by mouse bone marrow if given *in vivo* or *in vitro* (29); (b) prostaglandins; (c) interferon that produces a species-specific inhibition of granulocytic colony formation in addition to its antiviral effects (30); (d) hemopoietic inhibitory activity (HIA), closely related, if not identical to, tumor necrosis factor (TNF). This present report deals with studies relating to the activity (d), namely, hemopoietic inhibitory activity (HIA).

CSA is increased in mouse serum after infection, injection of endotoxin, and inoculation with BCG or *C. parvum*. Since TNF is produced in mice by a regimen involving the same immunologically active compounds, it was of interest to examine the TNF-rich serum as well as the partially purified TNF preparations for CSA activity.

METHODS

Young adult female CD-1 Swiss mice were injected intravenously with 1.0 mg (dry wt) of CP (Burroughs Wellcome) in 0.5 ml of sterile saline. Seven to 10 days later, 25 μg of *E. coli* endotoxin (Difco) was injected iv and 90-120 min later blood was collected by retroorbital punch. After determining that this serum was rich in TNF activity, it was used for the preparation of a partially purified fraction of TNF called G-200 II according to the method of Green *et al* (7). TNF was bio-assayed by determination of the extent of necrosis, which developed in 8-day-old intradermal Meth A tumors in BALB/c mice. Spontaneous regression of these tumors did not occur in untreated mice or in mice injected with normal mouse serum. Tumors in untreated mice were 100% lethal in 3-6 weeks. The degree of necrosis developing in the Meth A tumors, 24 hr after the iv injection of 0.5 ml of the test sample was graded as: no change (−); slight necrosis (+); moderate ne-crosis (++); or extensive necrosis (+++). Two mice were scored for assay of each sample. Assays were carried out in duplicate. As a standard, 0.5 ml of TNF-rich serum from CP-stimulated endotoxin-treated mice was injected iv which gave ne-crosis (+++) in the intradermal tumors. About 80% of the tumors that showed necrosis (+++) within 24 hr regressed completely in 2 to 3 weeks.

Assay for CSA Activity in Semisolid Agar

Test samples including normal mouse serum, TNF-rich serum, and G-200 II and CSA (10x conditioned medium of murine myelomonocytic cell line WEHI-3) (31) were assayed for granulocyte-macrophage colony growth in soft agar-bone marrow culture system. One-tenth ml of serial dilutions of test samples were assayed using 75,000 bone marrow cells from the femur of C57BL/6J mouse in 1 ml of culture medium per plate. The dilutions of TNF preparations were ex-pressed in terms of the amount of protein assayed.

Since TNF activity in serum samples was assayed in BALB/c mice, the TNF-containing samples were assayed for CSA using BALB/c mouse marrow as well. Since initial results indicated no major difference in the pattern of colony growth response between the two, all subsequent CSA assays were carried out using C57BL/6J mouse marrow cells according to the standard CSA assay.

Serum Inhibition of Normal and Neoplastic Cell Cloning

The assay system used here was identical to the assay for CSA, except that the samples to be tested for inhibition of colony growth were also plated at 0.1 ml with an exogenous source of 0.1 ml of CSA (10 x WEHI-3 CM) or its serial dilution per plate.

Cloning of L-cells, WEHI-3, and Meth A cells was carried out in semisolid agar culture in the absence of an exogenous source of CSA. Viable cells were determined by trypan-blue exclusion and plated at 1000 cells per milliliter per plate. Test samples for inhibition of cloning were assayed at 0.1 ml/plate. McCoy's 5A medium supplemented with vitamins, amino acids, sodium pyruvate, and L-glutamine with 15% fetal calf serum was used for all assays and the cultures were incubated at $37°C$ in humidified air and 8% CO_2 environment. The plates were scored at 7 days and a clone of 40 or more cells was considered as a colony.

RESULTS AND DISCUSSION

Sera from normal, CP-treated, endotoxin-treated, and CP-endotoxin-treated mice were assayed for CSA. Serum from endotoxin-treated mice showed the highest CSA activity with a plateau of activity at or above 875 μg serum protein. The plateau of CSA activity with serum from CP-treated mice was reached at 1750 μg of serum protein. The CSA activity of the normal mouse serum remained linear throughout the span of protein dilutions used and showed no evidence for the presence of inhibitory activity. Undiluted serum (7000 μg protein/plate) from CP-endotoxin-treated mice (TNF-rich serum) did not support colony growth. Upon serial dilution, CSA activity was noted and increased to maximum with increasing dilution to about 875 μg of serum protein. With further dilution, the CSA activity decreased.

The significance of the TNF in CP + endotoxin serum, which did not support the growth of bone marrow colonies, was examined. The partially purified TNF fraction (G-200 II) was assayed for CSA activity. As a control, a G-200 II fraction was prepared from normal mouse serum (n-G-200 II) and assayed. The TNF-positive G-200 II fraction (40 mg protein/ml) did not induce colony growth when used undiluted (4000 μg/plate); increasing numbers of colonies were observed with increasing dilution and CSA declined with further dilutions after reaching a plateau of activity (between 1000 μg - 500 μg protein/plate).

In contrast to whole normal serum, the n-G-200 II fraction failed to support colony growth, but gave increasing numbers of colonies with increasing dilution. This suggested the presence of an inhibitor of TNF in n-G-200 II. When bioassayed for TNF activity, n-G-200 II did not produce necrosis of tumor at 10 mg protein/mouse (the level for full activity with G-200 II) but, concentration of n-G-200 II and TNF testing at 80 mg protein/mouse was positive for induction of tumor necrosis. Thus TNF is present in low titers in normal mouse serum.

Fractionation of G-200 II on polyacrylamide gel electrophoresis (7) yields a homogenous fraction (G-200 II-B) containing all of the G-200 II TNF activity. When G-200 II-B was assayed for CSA, colony growth was not supported, indicating that TNF and CSA were not identical and that gel electrophoresis might have separated these two activities.

A G-200 II fraction was electrophoresed on polyacrylamide gel and four major areas of protein distribution, including G-200 II-B$_1$ were collected. These areas were designated A, A$_1$, B, C [top (−) to bottom (+)]; they were A = (5.8 mg protein/ml); A$_1$ = (2.1 mg protein/ml); B = (7 mg protein/ml); C = (3.5 mg protein/ml), respectively. To represent the distribution of CSA activity, each fraction was brought to 1 ml without regard to total protein concentration. Fractions A, A$_1$, and C gave varying degrees of CSA activity that increased to a maximum on dilutions and declined with further dilutions. The G-200 II-B fraction did not induce colony growth at all. When bioassayed for TNF, only the B fraction induced tumor necrosis. These results indicated that the suppression of colony formation observed with TNF-rich serum was due to TNF and that this activity existed concurrently with CSA in the partially purified G-200 II fraction.

In order to evaluate the ability of the TNF (G-200 II-B) to suppress CSA-induced granulocyte-macrophage colony growth, a constant amount of G-200 II-B was added to various dilutions of standard CSA preparations. A marked inhibition of colony growth resulted. Although this amount of TNF inhibited colony growth at low CSA levels, it did not significantly inhibit at maximal CSA activity. If one assumes that the suppression of CSA in G-200 II preparation was due to the TNF content, then the addition of purified TNF (G-200 II-B) to a preparation of G-200 II should further reduce the detectable CSA activity. In a typical experiment, CSA activities were assayed in G-200 II alone and in G-200 II to which purified TNF was added. We found all colony growth was suppressed where purified TNF was added.

To evaluate the dose-dependent inhibition of CSA by TNF, different concentrations of TNF (G-200 II-B) were added to an amount of CSA that induced maximum colony growth. These samples were assayed and the results showed a concentration-dependent inhibition of colony growth by TNF and produced 50% inhibition at 350 μg protein.

Studies were carried out to determine if the suppression of colony growth by TNF was due to inhibition of CSA action on the progenitor cells (CFU-c) or was a consequence of the direct cytotoxicity on CFU-c. Accordingly, bone marrow cells were preexposed to TNF by incubating for different periods of time and, after washing free of TNF, the cells were assayed in the presence of CSA. The results indicated that the TNF-preexposed colony growth was similar to the controls, and that TNF per se was not cytotoxic for CFU-c.

Whether the TNF and CSA are competitive and might be dependent on the time sequence with which they come in contact with the CFU-c surface is unknown. The possibility of a complex between the TNF and CSA that results in the inhibition of colony growth also existed. Bone marrow cells were preexposed to either TNF or CSA and were assayed without washing in the presence of added CSA or TNF, respectively. A preincubated mixture of TNF and CSA was also evaluated for its effect on bone marrow colony growth. The results suggest that the inhibition of colony growth by TNF in the presence of CSA is not competitive and that no interaction is taking place between the TNF and CSA.

Growth inhibition by TNF of L cells and Meth A tumor cells in liquid culture medium has been reported (6). Since many of the tumor cell lines could be cloned in semisolid agar, the inhibition of cloning of L cells, myelomonocytic leukemic WEHI-3 cells and Meth A tumor cells by TNF was evaluated. Cells were cloned in semisolid agar medium in the absence of exogenous CSA. We found a concentration-dependent inhibition of cloning in all of these cells and the complete inhibition of colony growth was achieved with G-200 II at a level of 400 μg protein/plate, while 70 μg of G-200 II-B completely inhibited the cloning of these same tumor cells. By extrapolation, 50% inhibition of cloning of these neoplastic cell lines corresponded to about 5 μg protein of TNF (G-200 II-B), about 70 times less than that required (350 μg) to produce 50% inhibition of bone marrow colony growth.

These investigations show the coexistance of CSA and TNF in the serum of CP-endotoxin-treated mice. Further, it has become apparent that these two biological activities are separate entities and that the inhibition of colony growth by purified TNF (G-200 II-B) was specific and was not due to the glycoprotein nature of TNF. Our findings of the presence of CSA-inhibitory activity in normal mouse serum (n-G-200 II) and the fact that a high concentration of this fraction does induce necrosis in Meth A tumors, raise the possiblity that a TNF-like material exists at low levels in normal blood and may play a regulatory role in marrow cell proliferation and function *in vivo*.

Furthermore, the marked increase in production of this inhibitory activity in *C. parvum*-endotoxin-treated mice suggests that in addition to its potential hemopoietic regulatory role, this activity may be a major mediator of nonspecific host antineoplastic defense mechanisms. This concept is supported by our observation that neoplastic cells are markedly more sensitive to growth inhibition by TNF than are normal hematopoietic stem cells.

References

1. Alexander, P., and Evans, R. *Nature New Biol.* **232**, 76, 1971.
2. Parr, I., Wheeler, E., and Alexander, P. *Brit. J. Cancer* **27**, 370, 1973.
3. Hibbs, J. B., Jr. *Science* **184**, 468, 1974.
4. Suter, E., Allman, G. E., and Hoffman, R. G. *Proc. Soc. Biol. Med.* **99**, 167, 1958.
5. Younger, J. S., and Stinebring, W. R. *Nature* **209**, 456, 1965.
6. Carswell, E. A., Old, L. J., Kassel, R. L., Green, S., Fiore, N., and Williamson, B. *Proc. Natl. Acad. Sci. USA* **72**, 3666, 1975.
7. Green, S., Dobrjansky, A., Carswell, E. A., Kassel, R. L., Old, L. J., Fiore, N., and Schwartz, M. K. *Proc. Natl. Acad. Sci. USA* **73**, 381, 1976.
8. Zacharius, R. M., Zell, T. E., Morrison, J. H., and Woodcock, J. J. *Anal. Biochem.* **30**, 148, 1969.
9. Helson, L., Green, S., Carswell, E., and Old, J. J. *Nature* **258**, 731, 1975.
10. Moore, M. A. S., and Williams, N. *J. Cell. Physiol.* **80**, 195, 1972.
11. Metcalf, D., and Moore, M. A. S. "Haemopoietic Cells." North-Holland, Amsterdam, 1971.

12. Stanley, E. R., and Metcalf, D. *Aust. J. Exp. Biol. Med. Sci.* **47**, 467, 1969.
13. Metcalf, D., and Stanley, E. R. *Aust. J. Exp. Biol. Med. Sci.* **47**, 453, 1969.
14. Van den Engh, G., and Rol, S. *Cell Tissue Kinet.* **8**, 579, 1975.
15. Metcalf, D., and Wahren, B. *Brit. Med. J.* **iii**, 99, 1968.
16. Metcalf, D. *Immunology* **20**, 727, 1971.
17. Quesenberry, P. J., Morley, A., Miller, M., Rickard, K., Howard, D., and Stohlman, F., Jr. *Blood* **41**, 391, 1973.
18. Sheridan, J. W., and Metcalf, D. *J. Cell. Physiol.* **80**, 129, 1972.
19. Sheridan, J. W., Metcalf, D., and Stanley, E. R. *J. Cell. Physiol.* **84**, 147, 1974.
20. Metcalf, D., MacDonald, H. R., and Chester, H. M. *Exp. Hematol.* **3**, 261, 1975.
21. Chan, S. H. *Aust. J. Exp. Biol. Med. Sci.* **49**, 553, 1971.
22. Granstrom, M. *Exp. Cell. Res.* **82**, 426, 1972.
23. Granstrom, M., Wahren, B., Gahrton, G., Killander, D., and Foley, G. E. *Int. J. Cancer* **10**, 482, 1972.
24. Granstrom, M. *Exp. Cell. Res.* **87**, 307, 1974.
25. Metcalf, D., and Russell, S. *Exp. Hematol.* **4**, 339, 1976.
26. Chan, S. H., and Metcalf, D. *Cell Tissue Kinet.* **6**, 185, 1973.
27. Beran, M. *Exp. Hematol.* **2**, 58, 1974.
28. Chan, S. H., Metcalf, D., and Stanley, E. R. *Brit. J. Hematol.* **20**, 329, 1971.
29. Metcalf, D. *Proc. Soc. Exp. Biol. Med.* **132**, 391, 1969.
30. McNeill, T. A., and Gresser, I. *Nature New Biol.* **244**, 173, 1973.
31. Ralph, P., Moore, M. A. S., and Nilsson, K. *J. Exp. Med.* **143**, 1528, 1976.

ROLE OF ADHERENT CELLS IN THE INDUCTION OF CYTOTOXIC T LYMPHOCYTES: PROMOTION, SUPPRESSION, AND ANTIGEN PRESENTATION

Gustavo Cudkowicz, Yee Pang Yung, and William W. Freimuth

The requirement of adherent macrophagelike cells for the primary induction *in vitro* of specific cytotoxic T lymphocytes was recognized as early as 1970 and reported by three groups of investigators during the 1971-1973 interval (1-3). It was then established that adherent cells played an indispensible but "accessory" role in the generation of effectors, distinguished from the "central" role of specific responder lymphocyte clones and antigenic stimulator cells. The accessory function could have been the release of factors or defined mediators into the culture medium resulting in polyclonal activation of responding T lymphocytes, activation of antigen-bearing cells, or both. Since adherent cells were also required for the *in vitro* induction of antibody responses (4, 5), and higher numbers of adherent cells resulted in suppression instead of promotion of cytotoxicity (1, 6), the activities of accessory cells were essentially viewed in the context of trophic influences on, or regulation of, lymphocyte functions. It was realized shortly afterward that gene products of the *I* region of the major histocompatibility complex (MHC) control interactions of adherent antigen-presenting cells with T lymphocytes in proliferative responses of cultured guinea pig and murine cells (7-9). The possibility was therefore raised that adherent cells might promote nonspecific as well as specific lymphocyte activation in the course of the generation of cytotoxic effectors.

Attempts to further purify and characterize adherent cell populations participating in the induction of cytotoxicity and to define discrete functions began in earnest in 1977 with the analysis of primary responses to allogeneic (10) or parental (11-13) stimulator cells, and to syngeneic cells modified by trinitrobenzene

209

sulfonate (14). These investigations confirmed and extended the earlier reports by demonstrating that (i) the competence for each response was lost after *in vitro* removal of adherent splenocytes by either the carbonyl iron plus magnetism or the Sephadex G-10 filtration techniques, and after the exposure of splenocytes *in vivo* and *in vitro* to antimacrophage agents; (ii) spleen cell preparations highly enriched with adherent, radioresistant, phagocytic cells, but depleted of T and B lymphocytes, were capable of restoring the competence for cytotoxic responses lost after the removal of adherent cells; (iii) radioresistant, adherent spleen cells exposed to ι-carrageenan were capable of inhibiting the lytic function of cyto-toxic T lymphocytes; (iv) purified adherent cells were excellent stimulators, in fact better stimulators, on a per cell basis, than other subpopulations of spleno-cytes, but interactions of adherent cells with lymphocytes were neither controlled (i.e., restricted) by the *I* region nor by other regions of the MHC in cytotoxic responses to allogeneic or modified syngeneic cells. Despite the considerable pro-gress in characterizing adherent cells (15, 16), in defining the requirements for adherent cells in cytotoxic systems (10-14), and in recognizing the antigen-pre-senting function of adherent cells in proliferative and humoral responses (17-19), still it has not been resolved whether adherent cells present antigen to prekiller T lymphocytes of specific clones, activate prekiller cells without regard to clonal distribution, or simply modulate lymphocyte activity via mediators, e.g., inter-ferons and prostaglandins.

We have addressed this issue by investigating the requirement for adherent cells of the primary F_1 hybrid antiparental *H-2^b* cell-mediated lympholysis (CML) response, and the genetic restrictions for promotion, suppression, and antigenic stimulation of the response by adherent cells. B6D2F_1 or B6C3F_1 hybrid respond-ing spleen cells were cocultured for five days with irradiated B6 stimulating spleno-cytes to generate specific antiparent cytotoxic T lymphocytes (CTL) (10, 21). The latter were assayed on lymphoma targets sharing with stimulator cells the homozygous MHC phenotype in a 4-hr chromium release test. This particular cyto-toxic response was selected because of its great sensitivity to antimacrophage agents (11, 12), genetic control by *H-2D* (21), and rather subtle autoreactive na-ture (22). Antiparent CTLs are capable of recognizing self-MHC products and of binding to syngeneic F_1 targets in direct cytotoxic tests. We expect that a response of this type is subject to regulatory controls and that the required adherent cells are the source of promoting as well as suppressive influences.

DEPRESSIVE EFFECTS OF ANTIMACROPHAGE AGENTS ON THE INDUCTION OF F_1 ANTIPARENT CTL

The role of macrophagelike cells in the generation of F_1 antiparent CTL was first investigated by means of silica particles (11) and three chemically distinct forms of carrageenans (12). These agents were selected because of preferential toxicity to macrophages; they were effective in abrogating or weakening the F_1 antiparent cell-mediated response upon *in vivo* administration to donors of responding cells or *in vitro* treatment of mixed spleen cell cultures.

The effectiveness of antimacrophage agents was dose dependent and varied with each agent; the active concentrations neither caused indiscriminate spleen cell death nor varied in relation to the presence of 2-mercaptoethanol in culture medium. Silica and carrageenans given *in vitro* had to be present in mixed spleen cultures during the inductive phase of CML, i.e., the initial 48 or 72 hr. During this interval, the exposure of responder and stimulator spleen cells to antimacrophage agents did not need to exceed 24 hr, but delayed administration of the agents, e.g., on the fourth or fifth day, had no inhibitory effect on the response. When the spleen cell treatment preceded stimulation (i.e., the mixing of responders and stimulators), it was sufficient to expose either the F_1 or the parental splenocytes to an antimacrophage agent to reduce or abolish the generation of CTL.

In vivo administration of silica or carrageenans to F_1 mice donating responder spleen cells had a long-lasting depressive effect on the competence for *in vitro* CML. These experiments represent another instance in which stimulator parental splenocytes were not treated and, thus, contributed viable macrophages to mixed cultures. Such cultures failed, nevertheless, to generate CTL.

Taken together, results of the experiments with silica and carrageenans suggest that viable macrophagelike cells of F_1 *and* parental origin were needed at the time when prekiller F_1 lymphocytes are induced to generate specific antiparent CTL. This interpretation rests on the assumption that the agents impaired primarily and selectively the function of macrophages, presumably via macrophage depletion. The available evidence on the biological effects of silica and carrageenans is reasonably supportive of this assumption, especially for relatively low concentrations of the agents, but two recent observations weaken the argument. First, carrageenans, but not silica, impair lymphocyte as well as macrophage functions *in vitro* at concentrations ten to twenty times higher than those required for the depression of F_1 antiparent CML (12). Second, adherent macrophagelike cells surviving exposure to carrageenans acquire the ability to inhibit the lytic function of CTL (13). The experiments with silica and carrageenans still implicate macrophagelike cells in the induction of F_1 antiparent CTL, but the evidence seems rather circumstantial. One could not exclude the possibility that the agents impaired directly or indirectly the functions of cells other than macrophages, and one could not clearly discriminate between the effect of macrophage depletion and activation of macrophage suppressors. The requirement of accessory cells for induction of CML should be confirmed by the more direct approach of adherent cell removal from splenocyte preparations and functional reconstitution with purified and characterized cells.

It is noteworthy that F_1 antiparent CML responsiveness was considerably more sensitive to antimacrophage agents than allogeneic CML responsiveness simultaneously tested (11, 12). According to data obtained with silica, the two cytotoxic responses were so different that one was fully abrogated (F_1 antiparent) while the other remained fully operational (F_1 antiallogeneic) under identical culture conditions, including the dose of silica. Substantive differences were detected also with carrageenans: exposure of responder F_1 cells to the agents for 24 hr before coculture with untreated stimulator cells resulted in loss of antiparent but not of antiallogeneic responsiveness. This apparent selectivity of antimacrophage agents for the depression of F_1 antiparent CML could not be attributed to the greater strength of allogeneic stimulation and cytotoxicity. The differential effect was still evident after the stronger response was weakened by adjusting responder-to-stimulator cell ratios for induction, and effector-to-target cell ratios for cytolysis. It is possible that the two responses differ in terms of accessory cell requirements: F_1 antiparent CML may be dependent on a larger pool of macrophagelike cells than allogeneic CML, so that incomplete depletion of such cells abolishes the former but not the latter response. It is likely, however, that the F_1 antiparent CML is distinctive with regard to mechanisms and accessory cell requirements in a qualitative rather than quantitative sense. Macrophagelike cells of parental origin (untreated stimulators) could not substitute for those of F_1 hybrid origin (treated responders) and *vice versa* whenever one of the cell preparations was exposed to antimacrophage agents. The opposite occurs in CML to allogeneic and modified syngeneic cells where adherent cells of responder and stimulator origin replace each other without noticeable restrictions (3, 10, 14).

DEPRESSIVE EFFECTS OF ADHERENT CELL REMOVAL ON THE INDUCTION OF F_1 ANTIPARENT CTL

Responder and stimulator splenocytes were either pretreated with carbonyl iron particles and magnetism or filtered through Sephadex G-10 columns to remove phagocytic and other adherent cells. There is no doubt that both procedures removed conventional macrophages since the frequency of cells ingesting latex particles decreased from ~7 to <1%. There is evidence, however, that the carbonyl iron procedure also removes nonphagocytic adherent cells (23) and that both procedures remove nonviable and broken cells (unpublished observations). As to the competence for F_1 antiparent CML of the depleted populations, results of two representative experiments are shown in Table 1. Treatment of the responder or stimulator population alone consistently resulted in the development of reduced albeit substantial cytotoxicity, but treatment of both populations entirely prevented the generation of CTL. The carbonyl iron procedure was somewhat inferior to Sephadex G-10 filtration, mainly because the effectiveness of adherent cell removal was more batch-dependent for carbonyl iron (Lymphocyte Separator Reagent, Technicon Instrument Co., Tarrytown, New York) than for Sephadex G-10. No attempts were made in these experiments to remove extensively adherent cells by repeated exposures to carbonyl iron or serial filtrations through Sephadex G-10. In fact, mixed allogeneic cultures of treated splenocytes

TABLE 1. Effect of Adherent Cell Removal from Responder and Stimulator Spleen Cell Populations on the Induction of F_1 Antiparent CTL

| Removal of adherent cells from[a] | | Percent specific lysis (%)[b] | |
| | | B6D2F$_1\alpha$B6 → L5MF-22 | |
Responders	Stimulators	Carbonyl iron treatment	Sephadex G-10 filtration
−	−	46.4	67.1
+	−	39.2	36.4
−	+	23.5	26.7
+	+	3.3	−0.4

[a] Aliquots of responder and stimulator spleen cells were either exposed to carbonyl iron and magnetism or filtered through Sephadex G-10 columns before culture. + or − indicates whether or not the cocultured cells were so treated.
[b] Equal numbers of B6D2F$_1$ and irradiated B6 spleen cells were cocultured for five days. Effectors were assayed on radio-labeled L5MF-22 lymphoma targets (H-2^b) at the effector-to-target ratio of 40.

were often not as incompetent as F_1-parent cultures, an indication that enough adherent cells were left behind to support allogeneic CML, the less demanding response with regard to accessory cells.

Removal of accessory cells was rendered more effective by prefiltration of splenocytes through nylon wool columns and selection of the carbonyl iron batch. Depletion of F_1 responder cells alone was then sufficient to abrogate antiparent responsiveness, despite the availability in culture of parental adherent cells (Table 2). The double procedure did not affect T-lymphocyte viability since the depleted F_1 cells were still capable of responding to nondepleted allogeneic instead of parental stimulators.

Removal of adherent cells from splenocytes by two different techniques fully confirmed the data previously obtained with antimacrophage agents. The generation of F_1 antiparent CTL is critically dependent on adherent cells of both responder and stimulator origin, presumably sensitive to the *in vivo* and *in vitro*

TABLE 2. Effect of Adherent Cell Removal from Responder Spleen Cell Populations on the Induction of F_1 Antiparent CTL by Nondepleted Stimulators

| Treatment of responder cells[a] | Percent specific lysis (%)[b] | |
	B6D2F$_1\alpha$B6 → L5MF-22	B6D2F$_1\alpha$C3H → 6C3HED
None	40.2	43.9
Nylon wool filtration	54.2	46.4
Nylon wool filtration plus carbonyl iron treatment	2.9	37.9

[a] Aliquots of responder spleen cells were filtered through nylon wool columns; aliquots of the nonadherent subpopulation were also exposed to carbonyl iron and magnetism.
[b] Equal numbers of responder and irradiated stimulator cells were cocultured for five days. F_1 antiparent effectors were assayed on L5MF-22 lymphoma targets (H-2^b), and F_1 anti-C3H effectors on cells of lymphoma 6C3HED (H-2^k).

toxicity of silica particles and carrageenans. The induction of F_1 antiparent CML and of certain other CML responses (e.g., against allogeneic or modified syngeneic cells) differ operationally with regard to at least two important parameters: (i) the extent of dependence on an intact pool of adherent cells (quantitative difference); and (ii) the extent to which the functions of adherent cells of responder and stimulator origin are interchangeable (qualitative difference).

FUNCTIONAL RECONSTITUTION WITH PLASTIC-ADHERENT CELLS OF SPLENOCYTES RENDERED INCOMPETENT BY SEPHADEX G-10 FILTRATION

Responder and stimulator spleen cells were rendered incompetent for F_1 antiparent CML by a single filtration through Sephadex G-10 columns and then co-cultured in the presence of added cells separated by adherence to plastic, according to Cowing et al. (15), and purified by irradiation (2000 rads of γ rays) and exposure to anti-Thy-1.2 antiserum plus rabbit C in vitro. 0.5 to 1.0 x 10^6 added cells were usually sufficient to fully restore F_1 antiparent (Table 3) as well as allogeneic responsiveness (data not shown). There were, however, definite genetic

TABLE 3. Restoration of F_1 Antiparent Responsiveness to Splenocytes Filtered through Sephadex G-10 Columns by a Plastic-Adherent, Radioresistant, Thy-1-Negative Subpopulation of Cells

Added cells[a]		Specific lysis (%)[b]	Restoration of
Strain	H-2	B6D2F$_1\alpha$B6 → L5MF-22	responsiveness[c]
None		−1.3	
B6	b	79.3[d]	+ +
C3H.B10	b	72.9	+ +
C3H	k	27.3	±
DBA/2	d	23.1	±
B6D2F$_1$	b/d	56.4[d]	+
B6C3F$_1$	b/k	39.0	±

[a] Responder and stimulator spleen cells were filtered through Sephadex G-10 columns. Added spleen cells were purified by adherence to plastic, irradiated in vitro (2000 rad of γ rays), and exposed to anti-Thy-1.2 antiserum plus rabbit C. 3 x 10^6 filtered responder, 3 x 10^6 filtered stimulator, and 7.5 x 10^5 adherent cells were cocultured for five days.
[b] F_1 antiparent effectors were assayed on radio-labeled L5MF-22 lymphoma targets (H-2^b) at the effector-to-target ratio of 40.
[c] + +, full restoration; +, partial restoration; ±, questionable restoration (presumably due to allogeneic stimulation of F_1 responders by added cells).
[d] Whereas reconstitution with B6D2F$_1$ adherent cells still required nonadherent B6 stimulator splenocytes in the culture for induction of antiparent CTL (accessory function), reconstitution with B6 adherent cells did not (accessory and stimulatory function).

restrictions for the restoration of F_1 antiparent CML. In an initial series of experiments, summarized in Table 3, the only adherent cells that fully restored B6D2F$_1$ anti-B6 CML were those sharing the $H\text{-}2^b$ haplotype of B6 stimulator splenocytes, irrespective of the genetic background. Adherent cells sharing the $H\text{-}2^{b/d}$ haplotypes of F_1 responder cells usually afforded partial restoration; cytotoxic activity of cultures so restored varied considerably from one experiment to another, probably in relation to the number of residual adherent cells of parental origin. Adherent cells bearing the homozygous $H\text{-}2^k$ or $H\text{-}2^d$ haplotypes that are allogeneic to B6 stimulators, or the heterozygous $H\text{-}2^{b/k}$ haplotype combination that is semiallogeneic to F_1 responder cells, reconstituted marginally, if at all. The low values of parental target cell lysis elicited by effectors generated under the influence of allogeneic or semiallogeneic adherent cells are not attributable to reconstitution. Nonspecific triggering of antiparental F_1 lymphocyte clones can occur in the presence of strong allogeneic stimuli (unpublished observations) presumably mediated by diffusible factors.

In view of the MHC restriction for adherent cell function, it was necessary to determine whether adherent cells could serve as stimulators, accessory cells, or both. A series of experimental protocols outlined in Table 4 clarified this issue. Significant variables were the omission of parental splenocytes from Sephadex G-10 depleted or nondepleted cultures to which adherent cells of either F_1 or parental origin were added. The results were clear-cut. F_1 adherent cells served exclusively in an accessory capacity, as they were required for the induction of CTL but could not trigger F_1 responder cells in the absence of parental stimulators (see footnote d of Table 3). The accessory function was MHC restricted. Parental-strain adherent cells served both as accessory and stimulator cells and also these functions were MHC restricted.

THE POTENT STIMULATOR ACTIVITY OF PARENTAL $H\text{-}2D^b$ PLASTIC-ADHERENT CELLS IS INFLUENCED BY THE H-2K REGION

Titration of the stimulator activity of parental adherent cells for B6D2F$_1$ anti-B6 CML indicated that such cells are considerably more effective than nonseparated splenocytes. Comparable levels of specific cytotoxicity were induced by stimulating F_1 responders with 3.5×10^6 irradiated spleen cells or 4×10^5 irradiated plastic-adherent cells (Table 5). The activity of the latter population was even greater in the absence of contaminating T lymphocytes, irrespective of whether T-cell depletion was achieved via exposure of adherent cells to rabbit antimouse brain serum plus C, or congenital absence of the thymus from nude donor mice. 4×10^4 T-cell-depleted plastic-adherent cells were about as effective as 2×10^6 nonseparated splenocytes, a 50-fold enrichment of activity. These purified cell populations were also enriched for phagocytes ($\sim$70% Latex particle

TABLE 4. Experimental Protocols Designed to Distinguish between the Accessory and Stimulator Functions of Plastic-Adherent Spleen Cells

F_1 hybrid responders	Parental stimulators	Adherent cells (origin)	Function tested
G-10 filtered	G-10 filtered	F_1 hybrid	Accessory
G-10 filtered	None	F_1 hybrid	
G-10 filtered	G-10 filtered	Parental*	Accessory & Stimulator
G-10 filtered	None	Parental*	
Nonfiltered	None	Parental*	Stimulator

*Adherent cells with recombinant H-2^b haplotypes were employed for mapping of genetic restrictions.

ingesting cells), but the correlation does not constitute by itself a strong indication that phagocytic cells were the stimulators. The adherent cell preparations employed were still heterogeneous and could have contained dendritic (16, 18) as well as other cell types capable of stimulation.

The potency of plastic-adherent cells for inducing F_1 antiparent CML is appreciated not only from quantitative considerations but also from the ability of such cells to bypass the requirement of T lymphocytes for stimulation (24). Nonseparated parental spleen cells must contain T lymphocytes reactive against MHC alloantigens of F_1 responder cells *and* adherent cells to be effective as stimulators in this cytotoxic system. The significance of this observation is not yet understood.

Recombinant analysis was applied to the stimulator and accessory activities of H-2 homozygous plastic-adherent cells so as to map the restricting region or regions within the MHC. It should be kept in mind that the target structures for B6D2F_1 anti-B6 CML are controlled by H-$2D^b$. All of the protocols outlined in

TABLE 5. Potent Stimulator Activity of Purified B6 or B10 Adherent Spleen Cells for the Induction of B6D2F_1 Antiparent CTL

Stimulator cells (plastic-adherent and irradiated)[a]	Percent specific lysis with graded numbers of stimulator cells (%)[b]		
	4×10^5	1×10^5	4×10^4
B6	51.6^c	25.4	15.7
B6 (T-cell depleted)	58.3	45.4	27.0
B10.nu/+	52.6	9.7	4.1
B10.nu/nu	57.4	44.1	26.0

[a] Stimulator cells were purified by adherence to plastic, irradiated *in vitro* (2000 rad of γ rays), and aliquots were exposed to rabbit antimouse brain serum plus rabbit C. 3.5×10^6 nonseparated responder spleen cells and graded numbers of stimulator cells were cocultured for five days.

[b] F_1 antiparent effectors were assayed on radio-labeled L5MF-22 lymphoma targets at the effector-to-target ratio of 40.

[c] Comparable specific lysis values were obtained with 3.5×10^6 nonseparated B6 or B10 splenocytes.

Table 4 were employed for testing most of the informative recombinant $H\text{-}2^b$ haplotypes, and a summary of results is given in Table 6. First, it was confirmed that $H\text{-}2^b$ adherent splenocytes could serve as stimulators and accessory cells (group 1), whereas $H\text{-}2^d$, $H\text{-}2^k$, and $H\text{-}2^a$ cells could not do so (group 2). Genes other than those of the MHC, contributed by the background genome of inbred mouse strains unrelated to B6, had no detectable influence. It was sufficient for the b allele to be present at the $H\text{-}2D$ region to ensure function of adherent cells, provided that the other MHC regions had d alleles (group 3). The presence of b alleles at regions other than $H\text{-}2D$ did not ensure function of adherent cells (group 4). Surprisingly, however, $H\text{-}2D^b$-positive adherent cells failed to stimulate or otherwise promote the induction of cytotoxicity whenever k alleles were present at other MHC regions (group 5). Even when the K and IA regions only possessed k alleles, as for the $H\text{-}2D^b$-positive adherent cells of strain B10.A(4R), there was no function, an indication that one or both of these regions were restrictive. The constitution of the recombinant $H\text{-}2$ haplotype of strain B10.BYR being K^q, IA^k, D^b (group 6) is appropriate for resolving this issue. Since B10.BYR adherent

TABLE 6. Induction of B6D2F$_1$ Antiparent CTL by Plastic-Adherent Stimulator Cells Bearing Homozygous $H\text{-}2^b$ Recombinant Haplotypes

Group	Strains	K	A	B	J	E	C	S	G	D	F$_1$ antiparent response[b]
						Donors of adherent cells[a] — MHC regions					
1	B6 B10										+ +
	C3H.B10	b	b	b	b	b	b	b	b	b	+ +
	129										+ +
2	B10.D2 DBA/2	d	d	d	d	d	d	d	d	d	±
	C3H	k	k	k	k	k	k	k	k	k	±
	B10.A	k	k	k	k	k	d	d	d	d	±
3	B10.HTG HTG	d	d	d	d	d	d	d	.	b	+ +
	D2. GD	d	d	d	b	b	b	b	b	b	+ +
4	B10.A(5R)	b	b	b	k	k	d	d	d	d	±
	B10.A(18R) HTI	b	b	b	b	b	b	b	.	d	±
	B10.S(21R)	b	b	b	b	b	b	b	b	d	±
5	B10.AM	k	k	k	k	k	k	k	k	b	±
	B10.A(4R)	k	k	b	b	b	b	b	b	b	±
	B10.A(2R) HTH	k	k	k	k	k	d	d	.	b	±
6	B10.BYR	q	k	k	k	k	d	d	.	b	+ +

[a]Stimulator cells were purified by adherence to plastic and irradiated *in vitro* (2000 rad of γ rays). 10^6 of such cells were cocultured for five days with either 3.5×10^6 B6D2F$_1$ or B6D2F$_1$ plus B6 splenocytes, according to the protocols outlined in Table 4 and marked by an asterisk. The F$_1$ antiparent effectors were then assayed on radio-labeled L5MF-22 lymphoma targets at the effector-to-target ratio of 40.

[b]+ +, full response; ±, questionable response. The stimulator and the combined accessory plus stimulator functions of adherent cells were tested.

cells were functional, *H-2K* was identified as the restrictive region. Comparison of nonstimulatory B10.A(2R) and HTH cells (group 5) with the stimulatory B10.BYR adherent cells is especially compelling for reaching this conclusion. Thus, the functions of adherent cells in F_1 antiparent CML required H-$2D^b$, the allele controlling stimulator and target determinants (21), *and* a permissive allele at *H-2K*. K^b, K^d, and K^q are permissive, whereas K^k is nonpermissive. The data of Table 3 suggest that even heterozygosity for K^k, as in B6C3F$_1$ adherent cells ($K^{b/k}$), is nonpermissive. Mechanisms to be considered for this intergenic interaction are *H-2K*-linked immune response gene effects, influences by K^k on the expression of D^b products, and dual recognition, for which there are precedents in cell-mediated immune responses (25-29).

SUPPRESSION OF CTL BY CARRAGEENAN-ACTIVATED MACROPHAGELIKE CELLS

Suppressor cells capable of inhibiting the effector arm of F_1 antiparent and allogeneic CML are generated *in vitro* by nonstimulated spleen cells of normal mice cultured in the presence of sublethal concentrations of carrageenans (13). Cultured spleen cells not exposed to carrageenans fail to inhibit cytotoxic effectors.

The subpopulation of cells acted upon by carrageenans is present in the spleen of congenitally athymic nude mice and is moderately adherent to Sephadex G-10. It is not removed, however, by carbonyl iron and magnetism. Preexposure of splenocytes to radiation (2000 rads of γ rays) does not prevent the *in vitro* induction of suppressor cells by carrageenans; in short, this induction requires neither reproductive viability nor antigenic stimulation of spleen cells. The selective antimacrophage activity of carrageenans at low concentrations and the properties just described strongly suggest that suppressor function is simply activated in cells belonging to the monocyte-macrophage lineage. In fact, the matured suppressors remain radioresistant, become strongly adherent to Sephadex G-10, and lack surface Thy-1 antigen detectable by antibody plus C. Moreover, suppressor cell activity is neither restricted by the MHC nor by the immunologic specificity of CTL.

Inducibility of suppressor cells by macrophage-active agents and inhibition of the effector phase of CML are altogether not unusual (see ref. 13 for bibliography). BCG, *Corynebacterium parvum*, and M-locus stimulation have been employed to induce macrophagelike suppressor cells capable of inhibiting the inductive phase of CML, and the inhibition of mature CTL has been reported for T-lymphocyte suppressors. It is nevertheless of interest that macrophagelike cells can be readily activated to become inhibitors of the induction as well as the function of CTL, since nonspecific promotion and suppression may represent one of the roles of accessory cells in F_1 antiparent and other kinds of CML.

CONCLUDING REMARKS

There is no doubt that adherent spleen cell populations highly enriched with phagocytes participate in the induction of CTL. The requirement for such cells has been demonstrated in allogeneic, F_1 antiparent, and chemically modified syngeneic mixed spleen cell cultures during the initial period of responder-stimulator cell interaction. Adherent cells usually promote the induction of CTL, but under given conditions spleen cell subpopulations with similar properties inhibit the induction and function of CTL. In the CML responses against allogeneic and modified syngeneic cells, the promoting and suppressive activities of adherent cells are neither immunologically specific nor MHC restricted; the mechanism of accessory function by adherent cells may then be properly described as polyclonal lymphocyte activation via diffusible factors and/or defined mediators.

The situation is more complex for F_1 antiparent CML. First, the induction of this response is considerably more sensitive to adherent cell depletion than the induction of allogeneic CML; second, both responder and stimulator-type adherent cells seem to be required for full responsiveness, although the proportions need not be equal; third, the accessory and stimulator functions of adherent cells are restricted by the *H-2K* region; fourth, there are permissive and nonpermissive *H-2K* alleles so that adherent cell functions do not require *H-2K* identity with responder lymphocytes. Mechanisms of accessory function in this CML response may be multiple so as to include those operative in allogeneic CML and, in addition, specific (monoclonal?) lymphocyte activation, presumably via presentation of antigen. Whether antigen-presenting cells are phagocytic, dendritic, or otherwise in this system is yet to be determined.

Acknowledgments

This work was supported by U. S. Public Health Service research grants AM-13969 and CA-12844, National Institutes of Health, and by contract NO1-CM-53766, Division of Cancer Treatment, National Cancer Institute.

References

1. Lonai, P., and Feldman, M. *Immunology* **21**, 861, 1971.
2. Wagner, H., Feldmann, M., Boyle, W. and Schrader, J. W. *J. Exp. Med.* **136**, 331, 1972.
3. MacDonald, H. R., Phillips, R. A., and Miller, R. G. *J. Immunol.* **111**, 575, 1973.
4. Mosier, D. E. *Science* **158**, 1573, 1967.
5. Ly, I. A., and Mishell, R. I. *J. Immunol. Methods.* **5**, 239, 1974.

6. Miller, C. L., and Mishell, R. I. *J. Immunol.* **114**, 692, 1975.
7. Thomas, D. W., Clement, L., and Shevach, E. M. *Immunol. Rev.* **40**, 181, 1978.
8. Rosenthal, A. S. *Immunol. Rev.* **40**, 136, 1978.
9. Schartz, R. H., Yano, A. and Paul, W. E. *Immunol. Rev.* **40**, 153, 1978.
10. Davidson, W. F. *Immunol. Rev.* **35**, 261, 1977.
11. Shearer, G. M., Wakasal, H., Yung, Y. P. and Cudkowicz, G. *Cell. Immunol.* **39**, 61, 1978.
12. Yung, Y. P., and Cudkowicz, G. *J. Reticuloendothal. Soc.* **24**, 461, 1978.
13. Yung, Y. P., and Cudkowicz, G., *J. Immunol.* **121**, 1990, 1978.
14. Pettinelli, C. B., Schmitt-Verhulst, A. M. and Shearer, G. M. *J. Immunol.* **122**, 847, 1979.
15. Cowing, C., Schwartz, B. and Dickler, H. B. *J. Immunol.* **120**, 378, 1978.
16. Steinman, R. M., Kaplan, G., Witmer, M. D. and Cohn, A. Z. *J. Exp. Med.* **149**, 1, 1979.
17. Cowing, C., Pincus, S. H., Sachs, D. H. and Dickler, H. B. *J. Immunol.* **121**, 1680, 1978.
18. Steinman, R. M., and Witmer, M. D. *Proc. Natl. Acad. Sci. USA* **75**, 5132, 1978.
19. Singer, A., Cowing, C., Hathcock, K. S., Dickler, H. B. and Hodes, R. J. *J. Exp. Med.* **147**, 1611, 1978.
20. Shearer, G. M., Garbarino, C. A. and Cudkowicz, G. *J. Immunol.* **117**, 754, 1976.
21. Shearer, G. M., Cudkowicz, G. Schmitt-Verhulst, A. M., Rehn, T. G., Waksal, H. and Evans, P. D. *Cold Spring Harbor Symp. Quant. Biol.* **41**, 511, 1977.
22. Cudkowicz, G., Nakano, K. and Nakamura, I. Autoreactivity specific for murine antigens controlled by the H-2D region. *In* "Autoimmunity" (F. Milgrom and B. Albini, eds.), Karger Basel, Switzerland, 1979.
23. Golstein, P., and Blomberg, H. *Cell. Immunol.* **9**, 127, 1973.
24. Nakamura, I. and Cudkowicz, G. *Eur. J. Immunol.* 1979.
25. Schmitt-Verhulst, A. M., and Shearer, G. M. *J. Exp. Med.* **142**, 914, 1975.
26. Zinkernagel, R. M., Althage, A., Cooper, S., Kreeb, G., Klein, P. A., Sefton, B., Flaherty, L., Stimpfling, J., Shreffler, D. and Klein, J. *J. Exp. Med.* **148**, 592, 1978.
27. Doherty, P. C. Biddison, W. E., Bennink, J. R. and Knowles, B. B. *J. Exp. Med.* **148**, 534, 1978.
28. Kurrle, R., Röllinghoff, M. and Wagner, H. *Eur. J. Immunol.* **8**, 910, 1978.
29. Blanden, R. V., McKenzie, I. F. C., Kees, U., Melvold, R. W. and Kohn, H. I. *J. Exp. Med.* **146**, 869, 1977.

CYTOTOXICITY OF BONE MARROW MACROPHAGES FOR NORMAL AND MALIGNANT TARGETS

Marie-Louise Lohmann-Matthes

Studies by several independent groups have shown that macrophages may be activated and rendered cytotoxic for target cells by a product of sensitized T lymphocytes designated "macrophage cytotoxicity factor" (MCF) or "macrophage activating factor" (MAF) (1-3). We sought to determine at which stage of maturation macrophages could be rendered cytotoxic and whether neoplastic as well as normal cells could serve as targets for cytotoxic macrophages. Macrophages are required at an identical stage of maturation for these activities. A liquid culture system enabled us to cultivate macrophages from bone marrow precursor cells (4).

MATERIALS AND METHODS

Bone Marrow Culture Technique

Bone marrow cells of mice were cultured as previously described (4). Briefly, the cells were obtained from both femurs. After washing, they were seeded at $1\text{-}2 \times 10^6$ cells/12 ml bone marrow culture medium, using 10 cm Falcon plastic Petri dishes. The culture medium consisted of 50% Eagle's medium (Dulbecco modification), 15% fetal calf serum, 5% horse serum, and 30% conditioned medium. The conditioned medium was the 5-day supernatant of cultured embryonic fibroblasts or cultured L cells. These supernatants contain colony stimulating factor (CFS) (5). The bone marrow cell cultures were fed every third day. To isolate nonadherent cells, bone marrow cells were filtered through glass beads. The effluent cells, essentially devoid of phagocytes, were washed twice and suspended in culture medium.

Preparation of MCF

MCF was obtained by stimulating 10^8 mouse or human spleen cells with 10 μg Con A in 10 ml serum-free Eagle's medium. After 24 hr the supernatant was removed and filtered through 0.2 μm Millipore filters. Large batches of MCF-containing supernatants were pooled and titrated in terms of their ability to render macrophages cytotoxic.

Macrophage Activation and Cytotoxicity Test

At various intervals of bone marrow culture, cells were seeded into flat-bottomed glass tubes in 1 ml Eagle's medium containing 10% FCS. They were activated for 24 hr with an amount of MCF that had been shown to induce cytotoxicity, i.e., total lysis of DBA/2 5178Y target cells under appropriate conditions.

After the 24-hr activation period, the cultures were washed and ^{51}Cr-labeled targets were added. After an additional 18 hr, sediments and supernatants were separately counted. ^{51}Cr release was determined from the radioactivity recovered in the cell pellet and supernatant.

$$^{51}\text{Cr release} = \frac{\text{cpm supernatant}}{\text{cpm pellet} + \text{cpm supernatant}} \times 100$$

Specific ^{51}Cr release is taken as the difference in ^{51}Cr release between target cells cultured in the presence of activated macrophages (total release) or control macrophages (spontaneous release).

RESULTS AND DISCUSSION

Bone Marrow Cultures

The kinetics of bone marrow proliferation under the described conditions are shown in Fig. 1. The colony-stimulating factor (CSF) present in the conditioned

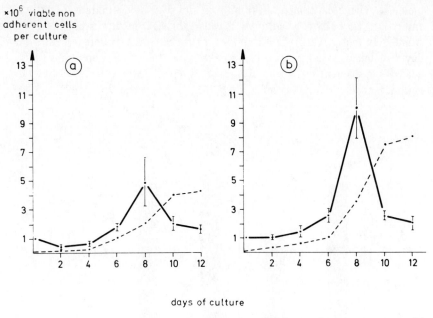

FIG. 1.   Kinetics of bone marrow cell proliferation. Triplicate cultures were set up with 10^6 cells. Solid line in (a) and (b) represent the average number of cells recoverable by gentle aspiration. Broken lines represent the number of adherent macrophages.

media induces proliferation and differentiation towards granulocytes and macro-phages. Beginning on day 3, all stages of maturation of both cell types are pre-sent. Proliferation of granulocytic elements terminates around day 5. At late stages of the cultures, cells of the macrophage series predominate. After 10-12 days of culture, a confluent monolayer of adherent macrophages is obtained. These macrophages are characterized by morphology, enzymes (esterase and phospatase), Fc receptors, capacity to kill intracellular bacteria, and capacity to lyse tumor cells (Domzig, W. *et al.* in preparation). The advantages of culturing macrophages from bone marrow are as follows: (a) cells can be harvested at any stage of maturation; (b) macrophages are not preactivated; (c) macrophages are obtained in large numbers from single donors.

Activation of Bone Marrow-Derived Macrophages by MCF

Both nonadherent supernatant cells and adherent macrophages can be acti-vated and rendered cytotoxic (Fig. 2). On day 5 of culture, nonadherent cells predominate, whereas on days 17-20 all cells were markedly adherent. The pos-

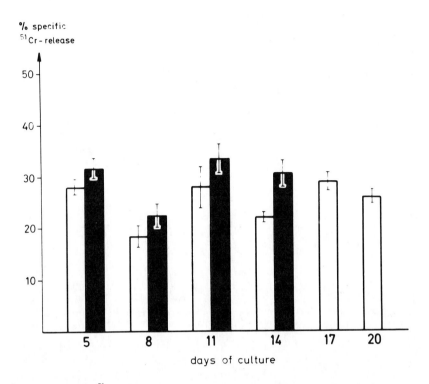

FIG. 2. Specific ^{51}Cr release from target cells exposed to activated bone marrow-derived cells. 3 x 10[5] cultured cells were incubated for 24 hr with MCF. E:T=12:1 in 18-hr cytotoxic assays. Spontaneous release in the presence of control effectors was 24%. ■, unseparated original bone marrow cells; □, nonadherent supernatant cells.

sible role of granulocytes as effectors has been ruled out before (4). Thus, not only the mature macrophages but also nonadherent precursors (e.g., promonocytes) can be activated to cytotoxic function.

Cytotoxicity of Macrophages Against Different Targets

The next set of experiments was undertaken to clarify the issue of selectivity in macrophage cytotoxicity against neoplastic versus normal targets (6-8). To compare these cytotoxic effects, macrophages of identical maturation and activation were employed. Eight-day-old bone marrow cultures provided the macrophages that were activated for 24 hr with 100 μl of a titrated MCF preparation.

TABLE 1. ^{51}Cr Release from Various Target Cells Incubated with Activated or Controlled Macrophages[a]

Target cells	Spontaneous release[b]	Total release[b]	Specific release
C57BL/E14	31 ± 2	70 ± 4	39 ± 3
DBA/2 5178Y	29 ± 2	71 ± 8	42 ± 5
SWISS A YAC	26 ± 3	76 ± 9	50 ± 7
3T3 SV40 (BALB/c)	35 ± 5	65 ± 7	30 ± 6
Meth C (BALB/c)	25 ± 2	40 ± 4	15 ± 3
3T3 (BALB/c)	35 ± 1	34 ± 2	—
BALB/c fibroblasts	32 ± 3	30 ± 2	—
DBA/2 lymphocytes	32 ± 2	42 ± 2	10 ± 3
DBA/2 Con A lymphoblasts	32 ± 2	51 ± 3	19 ± 3
Bone marrow macrophages, day 12	29 ± 2	30 ± 3	—
Bone marrow macrophage precursors day 5	30 ± 3	48 ± 5	18 ± 4

[a] The effector:target cell ratio was 8:1 in an 18-hr cytotoxic assay.

[b] Spontaneous release in the presence of control macrophages. Total release in the presence of activated macrophages.

Such macrophages kill a variety of tumor cells (in fact all tumors tested) after variable periods of incubation (Table 1).

Normal cells can also be attacked by activated macrophages. Nonneoplastic cells used as targets were either in the resting stage (e.g., mature macrophages, fibroblasts, lymphocytes) or mitotically stimulated (e.g., Con A blasts or bone marrow blasts of 5-day-old cultures). The data in Table 1 show that despite a preference for some tumor targets, lysis is also obtained with normal target cells, particularly after stimulation by mitogens. In contrast to this cytotoxicity, cytostasis by activated macrophages affects normal proliferating targets, such as bone marrow cells, to the same extent as tumor targets (Lohmann-Matthes, unpublished observation). Cytostasis is presumably a more relevant macrophage activity inducible by a variety of agents, whereas cytotoxicity is preferentially induced by lymphokines.

TABLE 2. Activated Human Macrophages and Mouse Macrophages Tested Against Human Target Cells[a]

Effector cells	^{51}Cr Release from melanoma RPMI 7932
Human control peripheral blood monocytes	25 ± 3
Human peripheral blood monocytes activated with 100 μl of human MCF	52 ± 4
Mouse control bone marrow macrophages	24 ± 2
Mouse bone marrow macrophages activated with 100 μl mouse MCF	70 ± 6

[a] Effector:target cell ratio=8:1 in an 18-hr cytotoxic assay.

Macrophage-Mediated Cytotoxicity in the Human System

Human peripheral blood monocytes, prepared according to Böyum (9), can be activated with human lymphokine and then lyse target cells (Table 2). The experiments in the human system are preliminary, but the results suggest similarities with the mouse system. Mouse macrophages are also cytotoxic for human target cells.

CONCLUDING STATEMENTS

Bone marrow cells cultured in liquid medium in the presence of CSF proliferate and differentiate along the granulocyte and macrophage pathways. The cells can be cultured in large quantities from one mouse and harvested for experiments at any stage of maturation and activation. They are never intentionally preactivated. These bone marrow-derived macrophages can then be activated to become cytotoxic by incubation with the lymphokine MCF. Both nonadherent precursor cells of early cultures as well as mature adherent macrophages of late cultures can be rendered cytotoxic. Thus, relatively immature macrophage precursors such as promonocytes can become cytotoxic even though the typical macrophage phenotype has not yet developed. The activated cells kill all tumor targets tested, but some cell types are more susceptible than others. The killing of target cells by activated macrophages is not restricted to neoplastic cells. Non-neoplastic cells are also susceptible to lysis, especially when they are mitotically stimulated.

References

1. Evans, R., Grant, C. K., Cox, H., Steele, K. and Alexander, P. *J. Exp. Med.* **136**, 1318, 1972.
2. Lohmann-Matthes, M.-L., Schipper, H. and Fischer, H. *Eur. J. Immunol.* **3**, 56, 1973.
3. Dimitriu, A., Dy, M., Thomson, N. and Hamburger, J. *J. Immunol.* **114**, 195, 1974.
4. Meerpohl, H. G., Lohmann-Matthes, M.-L. and Fischer, H. *Eur. J. Immunol.* **6**, 213, 1976.
5. Pluznik, D. H. and Sachs, L. *Exp. Cell Res.* **43**, 55, 1966.
6. Hibbs, J. B., Lambert, C. H. and Remington, J. S. *Proc. Soc. Exp. Biol. Med.* **139**, 1049, 1972.
7. Meltzer, M. S., Tucker, R. W., Sanford, K. K. and Leonard, E. *J. Natl. Cancer Inst.* **54**, 1177, 1975.
8. Holtermann, O. A., Klein, E., and Casle, G. P. *Cell. Immunol.* **9**, 339, 1973.
9. Böyum, A. *Scand. J. Clin. Lab. Invest. Suppl. 21,* 97, 77, 1968.

INTERFERON INDUCTION BY *CORYNEBACTERIUM PARVUM*

H. Kirchner, H. M. Hirt, and K. Munk

Corynebacterium parvum represents one of the strongest stimulants of the lymphoreticular system and has marked antitumor effects in certain experimental systems (for review see ref. 1). The lymphoreticular system plays a role in the defense against viral infections (2). However, the antiviral effects of *C. parvum* received little attention until recently when it was shown that mice could be protected against an experimental infection with herpes simplex virus by injection of *C. parvum* (3-5). Subsequently, we have demonstrated that spleen cells of mice injected with *C. parvum* elaborated *in vitro* an antiviral principle that was characterized as interferon (IF) (6).

IF, besides its antiviral effects, is known to have a number of so-called "side-effects" such as inhibition of tumor cell proliferation (7) and activation of macrophages (8, 9). It could well be that IF effects such as these are partially involved in the antitumor activity of *C. parvum*. Accordingly, we have further investigated IF production in mouse spleen cell cultures after stimulation with *C. parvum*.

MATERIALS AND METHODS

C57BL/6/J BOM (B6) mice, C57BL/6/J BOM/nu/nu mice and the heterozygous litter mates were obtained from Bomholtgard. Formalin-killed *C. parvum* (strain CN 6134), kindly provided by Dr. M. T. Scott (Burroughs Wellcome) was injected into B6 mice intraperitoneally (ip) at a dose of 350 µg. In other experiments, spleen cells of previously untreated mice were used that were treated *in vitro* with *C. parvum* organisms washed three times.

227

Single spleen cell suspensions were cultured in medium RPMI-1640 supplemented with 5% fetal bovine serum, antibiotics, and glutamine in 75 ml tissue culture flasks (Falcon) at a cell density of 3×10^6/ml. After various times of incubation, the cells were removed by centrifugation and the cell-free supernatants were recovered and stored at $-70°$C.

Treatment of spleen cells by antitheta serum plus complement and removal of B lymphocytes by nylon wool columns was performed as described in (10) and (11), respectively.

The capacity of IF-containing supernatants to inhibit the replication of vesicular stomatitis virus (VSV) in L cells was tested. Confluent monolayers of these cells in Leighton tubes (Bellco) were preincubated with the supernatants or with IF standards for 18 hr, thoroughly washed, and infected with VSV at a multiplicity of infection of ten. After one cycle of virus replication (15-18 hr), the tubes were frozen and thawed and the titer of VSV in the cell-free supernatant was determined by a viral plaque assay. Standard curves were established using international reference IF (G 001-904-511, NIAID, NIH) and the titers of IF of each supernatant were calculated as IU/ml.

The lymphocyte stimulation assay and the assay of growth inhibition of tumor cells (GIA) were performed as described in references (11) and (12).

RESULTS

Spleen cells from B6 mice were cultured 5-15 days after injection of *C. parvum* and produced IF (Table 1). IF was also produced when spleen cells from normal mice were challenged with *C. parvum in vitro*. The highest titers were produced when spleen cells of *C. parvum*-treated mice were rechallenged with *C. parvum in vitro*.

Spleen cells from nude mice produced the same amounts of IF as spleen cells from the heterozygous controls (Table 2). Treatment of B6 spleen cells with anti-Thy-1 serum plus complement did not abrogate *C. parvum*-induced IF production

TABLE 1. Production of IF by Murine Spleen Cells after Various Treatments with *C. parvum*

Exp.	Treatment by *C. parvum*[a]			Untreated control
	in vivo	*in vitro*	*in vivo* and *in vitro*	
A	310[b]	425	915	<10
B	560	390	610	<10
C	715	640	825	<10

[a] C57BL/6 mice were injected ip with 350 μg *C. parvum* and 10 days later their spleens were removed and single cell suspensions prepared. These were cultivated with or without 35 μg/ml *C. parvum* for 24 hr. Spleen cells from untreated mice similarly were cultivated with or without *C. parvum*.

[b] IU of interferon per milliliter recovered from spleen cells after 24 hr of cultivation.

TABLE 2. Induction of IF in Spleen Cell Cultures of Nude Mice by *C. parvum*

Source of spleen cells	Treatment	
	Without *C. parvum*	35 µg/ml *C. parvum*
nu/nu mouse # 1	$<10^a$	270
# 2	<10	130
# 3	<10	420
nu/+ mouse # 1	<10	830
# 2	<10	260
# 3	<10	320

[a] IU of interferon per milliliter.

(Table 3), but removal of B lymphocytes by nylon wool columns did abolish IF production. The latter maneuver, however, also removes many macrophages, and we have previously shown that macrophages are required for *C. parvum*-induced production of IF (13). Thus, it was important that nylon wool purified spleen cells, as opposed to spleen cell cultures freed of macrophages (but not of B cells) by plastic adherence, could not be restored by peritoneal exudate macrophages. Collectively, these data suggest that in mouse spleen cell cultures the cells induced by *C. parvum* to produce IF were B lymphocytes. This interpretation is further supported by the recent demonstration that *C. parvum* is a B cell mitogen for mouse spleen cells in culture (13, 14).

Supernatants of *C. parvum*-treated spleen cell cultures were extensively dialyzed against distilled water, 20-fold concentrated, and sterilized by filtration. Such supernatants had a marked inhibitory effect on mitogen-induced lymphocyte stimulation and on the *in vitro* proliferation of RBL-5 lymphoma cells (Table 4). Low titers of IF were also seen in concentrated control supernatants and these were also somewhat inhibitory. However, the differences between the two types of supernatants were highly significant in a large number of experiments.

Dialyzed and concentrated supernatants of *C. parvum*-treated spleen cells were also found to be active in the GIA (Table 5). When RBL-5 were cocultured with PEC at a low effector cell target cell ratio, no inhibition of the proliferation of RBL-5 was observed. In fact, in many experiments PEC had a growth-promoting

TABLE 3. Effect of Treatment with Anti-Θ Serum and Filtration Through Nylon Wool Columns on the Production of IF by Spleen Cells

Exp. no.	Treatment	Interferon titer
I	None	875^a
	Complement alone	790
	Anti-Θ + complement	815
II	None	635
	Nylon wool column	<10
	Nylon wool column reconstituted with 5% PEC	<10

[a] IU of interferon produced after 1 day of incubation.

TABLE 4. Inhibition of Proliferative Activity by Supernatants of *C. parvum*-Treated Spleen Cell Cultures

Tissue culture	Mitogen	Type of supernatant added:		
		A^a (<10 IU IF/ml)	B^a (44 IU IF/ml)	C^a (520 IU IF/ml)
RBL-5 (2 x 10^5 cells/ml)	—	220,861^b	78,532	12,411
B6 Spleen Cells (3 x 10^6 cells/ml)	—	954	863	421
B6 Spleen Cells (3 x 10^6 cells/ml)	PHA (1.25 µg/ml)	181,743	143,873	63,587
B6 Spleen Cells (3 x 10^6 cells/ml)	Con A (1.25 µg/ml)	247,758	198,694	97,688
B6 Spleen Cells (3 x 10^6 cells/ml)	LPS (10 µg/ml)	85,391	38,281	12,114

[a] A = preaged medium (propagated without spleen cells), B = supernatant of unstimulated spleen cells, C = supernatant of *C. parvum*-treated spleen cell culture. All three were extensively dialyzed against double distilled water, concentrated 20-fold and sterilized by filtration. 0.1 ml was added per ml of tissue culture.

[b] Mean cpm of triplicate cultures, determined by TdR-^{3}H incorporation between 44 and 48 hr after initiation of the cultures.

TABLE 5. Activation of PEC for Growth Inhibition Promoted by *C. parvum*-Induced Spleen Cell Supernatants

	Type of supernatant		
	A^a	B	C
PEC^b	984^b	893	914
RBL-5^c	241,931	171,445	120,085
PEC + RBL-5	328,351	234,483	15,148

^{a} For explanation see Table 4. Only 0.01 ml of the supernatants/ml of culture was added.
^{b} Cell concentration: PEC 1×10^6 cells/ml; RBL-5 2.5×10^5 cells/ml.
^{c} RBL-5 is a Rauscher virus-induced lymphoma cell line syngeneic to C57BL/6 mice (ref. 12).

effect. When the supernatants were added at a concentration at which they had less antiproliferative effects by themselves, a marked inhibition by PEC of the proliferation of RBL-5 cells was observed. Control supernatants had no effect in this assay. It is tentatively concluded that the IF-containing supernatants had an activating effect on the PEC.

DISCUSSION

Spleen cells of mice injected with *C. parvum* have been shown to inhibit the growth of lymphoma cells *in vitro* and also to suppress the proliferation of mitogenically activated lymphocytes (15). The effector cells in both cases appear to be macrophages. Our studies have now shown that IF is produced in cultures of mouse spleen cells after exposure to *C. parvum*. Inasmuch as IF has been shown to have an antiproliferative effect both on tumor cells and on mitogen activated lymphocytes, one can speculate that IF mediates the antiproliferate effect of macrophages. However, as of now our data do not support the concept that IF is a product of macrophages in *C. parvum*-stimulated spleen cell cultures.

We have observed two effects of IF-containing supernatants from *C. parvum*-stimulated cultures. One was a direct antiproliferative effect on tumor cells and lymphocytes; this is not surprising in view of similar data recently reported for other types of IF. Type I IF has been used in such studies, while the *C. parvum*-induced IF represents type II (manuscript in preparation). However, to demonstrate the antiproliferative effect of *C. parvum*-induced IF, the supernatants had to be dialzyed and concentrated. This makes it doubtful whether such a direct effect plays a role in the primary system when spleen cells from *C. parvum*-treated mice are cultured with tumor cells.

The other effect we have observed—the activation of PEC by *C. parvum*-induced IF-containing supernatants—could be demonstrated with lower concentrations of IF and was also seen when untreated supernatants were tested (data not shown). These data suggest that in *C. parvum*-stimulated cell cultures a two-step mechanism takes place: first, induction of IF by cells other than macrophages, presumably B cells, and then activation of macrophages by IF.

The nonantiviral effects of IF have always been viewed with skepticism since most IF preparations are relatively crude. It has, therefore, been difficult to exclude the possibility that effects such as those described were caused by contaminants in the preparations. For some of these IF activities, fairly good evidence has been presented that they are due to IF itself. In our experimental model it is not possible at present to resolve this basic issue. While it is well established that IF can be antiproliferative (7) and activate macrophages (9), and while we were able to demonstrate these activities with IF-containing supernatants, there is still the possibility that other "factors" in the supernatant, such as "macrophage activating factor" (16, 17), are responsible for them.

References

1. Scott, M. T. *Semin. Oncol.* **1**, 367, 1974.
2. Allison, A. C. *Progr. Med. Virol.* **18**, 15, 1974.
3. Kirchner, H., Hirt, H. M., and Munk, K. *Infect. Immun.* **16**, 9, 1977.
4. Glasgow, L. A., Fischbach, J., Bryant, S. M., and Kern, E. A. *J. Inf. Dis.* **135**, 763, 1977.
5. Morahan, P. S., Kern, E. A., and Glasgow, L. A. *Proc. Soc. Exp. Biol. Med.* **154**, 615, 1977.
6. Kirchner, H., Hirt, H. M., Becker, H., and Munk, K. *Cell Immunol.* **31**, 172, 1977.
7. Gresser, I., Brouty-Boyé, D., Thomas, M. T., and Macieira-Coeltro, A. *Proc. Natl. Acad. Sci. USA* **66**, 1052, 1970.
8. Huang, K. Y., Donahoe, R. M., Gordon, F. B., and Dressler, H. R. *Infect. Immun.* **4**, 581, 1971.
9. Schultz, R. M., Papamatheakis, J. D., and Chirigos, M. A. *Science* **197**, 674, 1977.
10. Holden, H. T., Kirchner, H., and Herberman, R. B. *J. Immunol.* **115**, 327, 1975.
11. Kirchner, H., Hirt, H. M., Kleinicke, CH., and Munk, K. *J. Immunol.* **117**, 1753, 1976.
12. Kirchner, H., Holden, H. T., and Herberman, R. B. *J. Natl. Cancer Inst.* **55**, 971, 1975.
13. Hirt, H. M., Kochen, M., and Kirchner, H. *Eur. Reticuloehdothelial Soc. Symp. Macrophages Cancer*, Edinburgh, 1977.
14. Zola, H. *Clin. Exp. Immunol.* **22**, 514, 1975.
15. Kirchner, H., Holden, H. T., and Herberman, R. B. *J. Immunol.* **115**, 1212, 1975.
16. Evans, R., and Alexander, P. *Transplantation* **12**, 227, 1971.
17. Lohmann-Matthes, M. L., Ziegler, F. G., and Fischer, H. *Eur. J. Immunol.* **3**, 56, 1973.

RECAPITULATION AND ASSESSMENT OF THE CONFERENCE DISCUSSIONS*

Gert Riethmüller, Gustavo Cudkowicz, and Maurice Landy

One of the latest developments in identifying host defense systems was the recognition that cytotoxic cells for a variety of targets occur naturally in the lymphoid system of mice, man, and other mammals (NK cells). An analogously broad system of cells was already known whose main feature was the requirement for antibody-coated targets (K cells). Both systems are constituted of cells that pose difficulties in categorization, as they do not express in the usual way cell surface structures regarded as markers for the two major categories of lymphocytes. Another common feature of these naturally occurring cells is their heterogeneity expressed as effector activities for given targets and requisite conditions for lysis. Both kinds of cells are cytotoxic and cytolytic and are generally detected by release of membrane-bound radioisotopes from target cells. Although selective, such cytotoxicity is not MHC restricted. The emergence of natural cell-mediated resistance brings to new prominence the older considerable body of information on natural antibodies. It bears emphasis that these constitute a wide spectrum of individual specificities whose elicitation has never been established unequivocally. In a way, the situation for NK cells is strikingly analogous in that they too, while noninduced, display distinct "selectivities." For both natural antibodies and natural killer cells, the eliciting mechanism is quite

---

* It was decided to replace the actual detailed discussion that followed most of the oral presentations at the conference with this recapitulation inasmuch as the subsequently submitted manuscripts in many instances had utilized discussion material. In the process of preparing this overview, it was almost inevitable that a measure of assessment and projection were also included. As this statement was prepared at a time well after the conference, it was also possible to include recent new data relevant to several of the major issues.

unknown; neither natural antibodies nor NK cells are discernible in the neonate, but both emerge after the first weeks of life.

If natural antibody is to confer the means of recognition, either by coating the targets or arming the effectors, there already exists an ample basis for specificity. Should NK cells prove to function without involvement of antibody, then these cells must utilize a novel, unfamiliar type of recognition structure.

FUNCTIONAL ROLE OF NK CELLS

Since their discovery, natural cytotoxic cells have attracted considerable attention as it was early perceived that they could fulfill major functions in the maintenance of homeostasis. They could exert surveillance against neoplastic transformation, and they could also regulate normal cell proliferation and differentiation in systems involving cell renewal. As to the surveillance function, it was early apparent, and continues so, that transformed cells are particularly susceptible targets—among them tumors of various histologic types, virus-transformed cells, and cells infected with intracellular pathogens, even protozoa. Indeed, target cells syngeneic with the effectors are susceptible to NK lysis as this presumably does not involve determinants coded for by histocompatibility genes.

The NK element has revived the surveillance concept as a non-T cell phenomenon at the very time of its apparent demise as a T-cell dependent system. Since athymic mice are fully competent with respect to NK activity, there is provided a basis for their unimpaired tumor resistance and, in fact, for their survival free of microbial infection as well as tumors.

In considering a general regulatory role for K and NK cells, one major obstacle has been lack of evidence for NK activity against normal targets. However, this could be more apparent than real (1-6), and suppressive influences (1,4,6) could mask such activities, especially as attention has, up to now, been focused on cytolysis tests rather than cytostasis. The fact is, that thymocytes, hemopoietic stem cells, peritoneal cells, and fibroblasts are susceptible to NK activity, indicating that normal cells can also be influenced.

Multiple correlations have been discussed between NK activity as measured *in vitro* on YAC-1 lymphoma targets, and natural resistance to hemopoietic cells as measured *in vivo* (6). These correlates include thymus independence of effector cells, their spontaneous occurrence in mice during the fourth week of life, and a paucity of conventional lymphoid cell markers. These correlates suggest that the mediators of resistance to hemopoietic grafts are a subpopulation of NK cells endowed with "selectivity" for determinants expressed on hemopoietic cells. Findings such as these provide a basis for the postulate that the NK system does in fact modulate hemopoiesis.

NK VERSUS K CELL DISTINCTION

There was considerable discussion of the analogies and distinctions that could be drawn between NK and K cells. The preponderance of evidence was against clear-cut distinctions: a number of correlations were pointed out in mouse and man with respect to age of maturation, organ distribution, and response to external agents. Moreover, blocking experiments with cold inhibitor cells indicated that there was cross inhibition between NK and K functions (7), but the issue is far from being resolved (8-10). More important, however, was the evidence brought forward in support of *in vivo* arming in human systems. Trypsin-treated effector cells regained specific activity upon incubation with antibody eluted from targets. This *seemed* to constitute a convincing example of *in vitro* arming. The fact is, however, that this apparently straightforward approach has not been reproduced. The issue has been readdressed by using anti-Ig reagents (purified Fab fragments) to inhibit cytotoxic activity. Here too, the inhibitory effect of these reagents implied that Ig was involved and functioned as the recognition structure. It should be noted that the inhibiting activity of anti-Ig reagents was not confirmed. Moreover, this inhibitory effect was not obtained in the most critical situations, i.e., where effectors and targets were of autologous origin.

NATURE OF EFFECTOR CELLS

There was a consensus that NK cells bear Fc receptors, but a distinction was made concerning its high and low expression and its class specificity. A functional role for this type of receptor was affirmed by the loss of cytotoxicity upon exposure to Ag-Ab complexes or aggregated IgG. Such evidence is considerably short of a formal demonstration that the Fc receptor functions by way of being the acceptor for the arming or target-coating antibody.

During the conference, there was an extensive debate on the identity of NK cells. The sharpest issue was whether marker of T cells are present on the surface of NK effectors. The general opinion was that in the mouse the Thy-1 marker is not readily detectable. This was taken to mean that if NK cells were to belong to the T-cell lineage, they would belong to the prethymic pool of cells expressing less Thy-1 that would be demonstrable by killing techniques. In the human system the situation is more complex. The existing evidence is as follows: NK activity is recovered among both low- and high-affinity E-rosettes and also among the nonrosetting lymphocytes. In fact, when the activity is expressed in lytic units, there is either more or as much activity among nonrosetting as there is among rosetting subpopulations. An anti-T cell serum plus C that inhibited E-rosette formation failed to affect NK activity. This type of data is inadequate for concluding that NK cells can be classified as T lymphocytes. Rather, it appears that the rosetting approach cannot separate NK-active from inactive subpopulations of peripheral blood leukocytes. Instead, NK effectors in man seem to represent a continuum of cell types extending across the conventional compartmentalization into classes of T and non-T lymphocytes (11). A noteworthy development in cellular immunology has been the gradual reassignment of markers, formerly thought to be restricted to one of the major lymphocyte classes, to now include the other. A consequence has been the considerable blurring of what had initially been an overly sharp distinction. NK cells may well be an example of a subpopulation of effectors that defies such simple categorization.

Still another approach to ascertaining the nature of NK cells is to examine their activity in various defined immunodeficiency states (12, 13). While severe combined immunodeficiency disease in which there is a stem cell defect entails loss of NK activity, the more selective immunodeficiency disorders, such as X-chromosome linked agammaglobulinemia and common variable agammaglobulinemia, leave NK activity essentially unaffected. Likewise, even the T-cell-deficient state of thymic aplasia (De George's syndrome) fails to lower NK activity. Thus, the experiments of nature serve to affirm the impossibility of assigning NK cells to any of the recognized lymphocyte classes.

Although NK cells are found exclusively in peripheral blood and spleen, the question of their lymphocytic nature has not been satisfactorily resolved. Those considering them as lymphoid cells have advanced as supporting evidence their distribution, the presence on their surface of Fc and E receptors, and, in the mouse, the presence of Thy-1 antigen. Moreover, the fact that they are nonadherent and nonphagocytic, and that they are present in peripheral blood, also has been taken to indicate, by way of exclusion, that NK cells are lymphoid cells. These "arguments" in favor of the lymphocytic nature of NK cells are, however, countered by the fact that Fc receptors are also widely distributed among cells other than lymphocytes and that there is NK activity in nonrosetting cell populations. Finally, the evidence for Thy-1 antigen on NK cells was obtained under extenuating conditions and its demonstration was confined to cells from nude mice (14).

Despite the inadequacy of the criteria applied, there is still a consensus that NK and K cells do belong to the lymphoid system. We point out, however, that this stems primarily from the strong bias that a host response endowed with target selectivity and manifestly potential as a defense and/or regulatory mechanism could *only* be mediated by cells of the lymphatic system. The fact is, however, that this edifice rested solely on the absence of evidence to the contrary until target-effector interaction in the NK cell system were studied by transmission electron microscopy (15).

ONTOGENY AND REGULATION

The ontogeny of NK cells, apart from its inherent importance, is a meaningful approach in delineating the distinctive nature of this cell lineage. It turns out, however, that the available information is fragmentary and insufficient.

K and NK cells, along with the effectors of natural resistance against microbial invaders, mature relatively late in mice, i.e., in the fourth week of life, then require additional weeks to attain maximum activity. During the neonatal period, when NK activity is absent, suppressor activity can be demonstrated in mouse spleen (6). Such suppressor activity is manifested in T- and B-cell-mediated function as well as NK; different types of suppressor cells are involved.

In mice, as the lymphoid system matures, the demonstrable level of suppressive activity progressively declines—presumably balancing (at about 3-4 weeks) against the appearance of NK activity. This "balanced" relationship *seems* to be reciprocal—but presently, this is just an impression (6). A situation rather analogous to that in infant mice is seen in adults treated with the bone-seeking radioisotope ^{89}Sr (6) that selectively irradiates bone marrow and, thus, depletes it of hemopoietic cells. Here too, NK, K, and other natural resistance systems are impaired (16-18). It turns out that in this situation suppressor cells are just as heterogeneous (i.e., against T and B and NK cells) as those of infant mice, and become evident at the very time NK cell activity declines. It would be too easy to rationalize from these pieces of information that effector cells are a stable population, present throughout, and that the variable is the number and activity of the suppressor population. These wide fluctuations of suppressive activity may not only occur in ontogeny but also in response to the various agents that are known to influence NK activity. In other words, the action of such agents may prove to be limited to the suppressor populations (6, 19). Perhaps it is more realistic to view a second possibility according to which suppressor cells would influence differentiation and maturation of effector cells from precursors. In that event, the mature functional cells would not necessarily be present throughout.

In the two situations considered, the primary and critical parameter for the altered level of NK activity is the strength of the suppressor population. A further alternative is that the aforementioned variations in the level of suppressor activity are secondary events, in which case effector cells would fluctuate in response to external stimuli and suppressor cells would then follow in the opposite direction, always in a dynamic counterbalance. Resolution of this issue requires independent assays of NK and of suppressor function in purified and separated cell populations. Whereas such assays are available, meaningful cell separation has not been attained.

Despite the spontaneous development of both the K and NK systems, the level of their activity can be enhanced or depressed by any of a number of external stimuli. We have learned at this conference of the effects of BCG and of interferon in the human system. It is, moreover, well known that viruses, bacteria, and tumor cells can stimulate NK activity in mice and rats. This all points on the one hand to a substantial degree of flexibility and on the other to its firm control and regulation. Suppressor cells have been discerned as operative and presumably regulating this noninduced system just as they are known to do in the familiar T-cell-dependent cytotoxicity. The fact that such suppressors include more than one cell type (macrophages as well as nonphagocytic cells) is indicative of possibly more than a single mechanism by which regulation can be effected. In fact, all of the external influences on NK cell activity could be attributable either to direct action on the effector cells, on the suppressors, or both. Furthermore, the possibility has been raised that such external agents act indirectly on those cells by giving rise to potent mediators. The recent work on interferon in this regard is a most notable example, but there are reasons for believing that a number of other mediators (prostaglandins, lymphokines) could also be implicated.

There is an extensive literature on the striking effects of microbial components (cell wall peptidoglucans, bacterial endotoxins) on enhancement of natural resistance to bacterial and viral pathogens. Despite much work on mechanisms, these complex effects leading to enhanced or reduced resistance are still not fully understood. More recently these systems were extended to include altered resistance to transplanted tumors. In the light of current knowledge, it can easily be visualized that such microbial components rapidly and profoundly alter host resistance to infection via K and NK cells, their suppressor, and promoters.

Finally, it is noteworthy that the active moieties of cell wall of certain major bacterial genera (peptidoglucans) have recently been isolated, identified as muramyldipeptides, and fully synthesized (20). The availability of such water-soluble, low molecular weight, nontoxic, biologically active immunopotentiators (and their structural analogs) should greatly increase the prospects of identifying the responding cells.

RESTRICTED VERSUS NONRESTRICTED CTL AND AUTOREACTIVITY

An interesting hypothesis was presented that seeks to unify MHC-restricted and nonrestricted T-dependent cell-mediated cytotoxicity. The basis for this projection was the observation of overlap, at the clonal level, of these two kinds of responses. The suggestion was made that CTL monitor modified self components as a matter of course. This "normal" event is easily recognized provided that the self component being modified is highly polymorphic. Self components that are not polymorphic would not be perceived experimentally even when modified. Evidence for this unifying concept was presented and it was pointed out that 45%

of a relatively small sample of human cells shared a component modified by TNP. It was pointed out that T-dependent cell-mediated cytotoxicity seems to be readily induced *in vitro* against cells that fail to express serologically defined major components of the MHC, possibly with the participation of components of xenogeneic serum (28).

New information was presented showing that cell surface modifications relevant for restricted cell-mediated cytotoxicity did not require covalent binding. In fact, such modification was obtained simply by exposing lymphoid cells to soluble hapten-protein conjugates. This striking success stands in complete contrast to analogous work on foreignization of tumor cells for the purpose of immunotherapy, an area of major endeavors where successes have been few indeed.

As mentioned earlier, one would expect a cytotoxic system relevant for control mechanism to display some form of autoreactivity. Response to modified self meets this postulate for T-dependent cytotoxicity. A relevant example was provided by a report on cytotoxic responses directed against autologous leukemic blasts. Interestingly, the stimulation required the presence of foreign MHC products on third-party cells, but the target determinants were exclusively on autologous cells, as demonstrated by competitive inhibition tests. The requirement for third-party cells would be "difficult" to meet *in vivo*, one reason why responses against autochtonous tumor cells are generally ineffective. It thus appears that there is little likelihood of adoption of this model for therapy. Third-party cells may stimulate production of interferon via the effect of T-cell lymphokines.

NK activity could be viewed as a *noninduced* response to self components analogous in many ways to the autoreactivity *induced* by third-party cells and mediated by CTLs. There is, in fact, strong correlation between NK activity *in vitro* and resistance to syngeneic lymphomas *in vivo* (21). One example of induced autoreactivity by human peripheral blood lymphocytes was put forward at this conference and required third-party cells for its induction; the targets for CML were autologous leukemic blast cells. Additional examples of induced autoreactivity were subsequently reported (22-25). Again, a pool of third-party cells greatly increased the probability and efficiency of the autoreaction involving leukemic blast cells as targets (22), but other experimental manipulations were also effective (24, 25). Autoreactive mixed leukocyte reactions and cytotoxicity were obtained with nonneoplastic stimulators. Autoreactive cytotoxicity could also be induced by nonlymphoid stimulator cells such as fibroblasts (26). Parental T cells, however, induce or derepress autoreactive clones of F_1 hybrid splenocytes specific for target structures controlled by the MHC (25). What is noteworthy in all these situations is the ease of induction, provided that triggering cells are present as third-party or parental cells. The "triggering" event may actually be a release from suppression or activation via mediators such as interferon.

The autoreactive nature of NK activity and its independence of conventional inductive processes (i.e., clone amplification) could be viewed as being especially significant for homeostatic and surveillance functions, in that it meets the key requirement of prompt recognition of self. The *in vitro* effector function is defined for convenience as cytolytic activity or release of chromium from labeled target cells. *In vivo*, however, effector function need not result in target cell lysis, but rather in cytostasis. In this case the effect on target cells could be reversible, a quality requisite for any regulatory system.

MACROPHAGES

It is well known that macrophages influence the induction of CML in various ways. They can act as accessory cells during the early inductive phase, presumably by presenting antigen or contributing in some other way to lymphocyte activation. Macrophages can also become suppressor cells. Macrophage functions other than antigen presentation are not restricted by the MHC and operate equally well irrespective of their responder, stimulator, or third-party origin. It could well be that the nonrestricted macrophages provide essential factors for lymphocyte survival and function (monokines?). The recent demonstration of subpopulations of macrophages selectively expressing the products of different I subregions of the MHC (27) may enable the separation, on this basis, of subtypes with different functions. The latter approach may eliminate some of the confusion and overlap in this area of research.

The matter of direct macrophage cytotoxicity for tumor targets after "activation" by various immunologic and nonimmunologic procedures is controversial. It is often stated that such macrophages are selectively cytotoxic for tumors. The fact is that more detailed studies on an extended array of neoplastic and normal targets, including primary explants as well as recently established nonneoplastic cell lines, have made it clear that "normal" cells can also be susceptible to such effectors. For now it would seem more realistic to view this situation as reflecting major *quantitative* differences rather than a *qualitative* distinction between normal and neoplastic targets.

A preview of the future of this field, in the sense of a dissection of these complex systems at the molecular level, is offered by the elegant studies of Moore and collaborators on a number of macrophage influences via defined mediators, operating in a cascadelike sequence. The influences were exerted on hemopoietic, lymphoid, and NK cells, and the mediators considered at this conference were prostaglandins, tumor necrosis factor, colony-stimulating factor (CSF), and interferon. Thus, macrophages could represent a versatile and heterogeneous class of regulatory cells.

CODA

Subpopulations of NK cells, distinguishable by their selectivity in cytotoxic interactions with given targets, seem to function preferentially in certain defined sites, this despite their abundant presence in the peripheral circulation. This would imply that there is a site restriction for their activity, i.e., that they are a distinctive component of the so-called lympho-myeloid microenvironment, possibly integrated into a regulatory network.

What NK cells actually constitute is a reflection of operational criteria. When considering the nature of target cells, such as nonneoplastic hemopoietic cells, the function of NK cells is referred to as regulation of hemopoiesis. On the other hand, when considering the action of NK cells on lymphoma targets, NK activity is alluded to as surveillance. Finally, when the target cells are infected with intracellular pathogens, NK activity is viewed as antimicrobial defense. In these diverse situations, NK cells are manifestly capable of recognizing the altered state of the target. It is noteworthy that conventional receptors seen thus far are not involved in this kind of NK recognition, nor has it been possible to demonstrate the presence of any other structures that could function in this capacity. The conference viewed this impasse as *the* central issue. Either there exists another mechanism for recognition by lymphoid cells, not yet elucidated, or else NK cells are in a separate nonlymphoid category and operate via a truly novel recognition mechanism.

Acknowledgment

The preparation of this article was supported by the Jubilee Fund of the University of Tübingen, Germany and U. S. Public Health Service Grant CA-12844, National Cancer Institute, Bethesda, Maryland.

References

1. Parkman, R., and Rosen, F. S. *J. Exp. Med.* **144**, 1520-1530, 1976.
2. Herberman, R. B., Nunn, M. E., Holden, H. T., Staal, S. and Djeu, J. Y. *Int. J. Cancer* **19**, 555-564, 1977.
3. Nunn, M. E., Herberman, R. B., and Holden, H. T. *Int. J. Cancer* **20**, 381-387, 1977.
4. Osband, M., and Parkman, R. *J. Immunol.* **121**, 179-185, 1978.
5. Welsh, R. M., Zinkernagel, R. M., and Hallenbeck, L. A. *J. Immunol.* **122**, 475-481, 1979.
6. Cudkowicz, G., and Hochman, P. S. *Immunol. Rev.* **44**, 13-41, 1979.
7. Ojo, E., and Wigzell, H. *Scand. J. Immunol.* **7**, 297-306, 1978.
8. Koren, H., and Williams, M. S. *J. Immunol.* **121**, 1956-1960, 1978.
9. Santoni, A., Herberman, R. B., and Holden, H. T. *J. Natl. Cancer Inst.* **62**, 109-116, 1979.
10. Pollack, S. B., and Emmons, S. L. *J. Immunol.* **122**, 718-722, 1979.

11. Ozer, H., Strelkauskas, A. J., Callery, R. T., and Schlossman, S. F. *Eur. J. Immunol.* 1979, in press.
12. Koren, H. S., Amos, D. B., and Buckley, R. H. *J. Immunol.* **120**, 796-799, 1978.
13. Pross, H. F., Gupta, S., Good, R. A., and Baines, M. G., *Cell. Immunol.* **43**, 160-175, 1979.
14. Herberman, R. B., Nunn, M. E., and Holden, H. T. *J. Immunol.* **121**, 304-309, 1978.
15. Roder, J. C., Kiessling, R., Biberfeld, P., and Andersson, B. *J. Immunol.* **121**, 2509-2517, 1978.
16. Haller, O., and Wigzell, H. *J. Immunol.* **118**, 1503-1506, 1977.
17. Bennett, M., Baker, E. E., Eastcott, J. W., Kumar, V., and Yonkosky, D. *J. Reticuloendothel. Soc.* **20**, 71-87, 1976.
18. Bennett, M., and Baker, E. E. *Cell. Immunol.* **33**, 203-210, 1977.
19. Cudkowicz, G., and Hochman, P. S. *In* "Developmental Immunobiology—Fifth Irwin Strasburger Memorial Seminar on Immunology," (G. W. Siskind, ed.), pp. 1-24. Grune and Stratton, New York, 1979.
20. Leclerc, C., Löwy, I., and Chedid, L. *Cell. Immunol.* **38**, 286-293, 1978.
21. Haller, O., Hansson, M., Kiessling, R., and Wigzell, H. *Nature* **270**, 609-611, 1977.
22. Zarling, J. M., Robins, H. I., Raich, P. C., Bach, F. H., and Bach, M. L. *Nature* **274**, 269-271, 1978.
23. Van de Stouwe, R. A., Kunkel, M. G., Halper, G. P., and Wecksler, M. E. *J. Exp. Med.* **146**, 1809-1814, 1977.
24. Miller, R. A., and Kaplan, H. S. *J. Immunol.* **121**, 2165-2167, 1978.
25. Cudkowicz, G., Nakano, K., and Nakamura, I. *In* "Autoimmunity, Proceedings of the 6th International Convocation on Immunology." Milgrom, F., and Albini, B., eds., S. Karger AG, Basel, 1979.
26. Ilfeld, D., Carnaud, D., and Klein, E. *Immunogenetics* **2**, 231-240, 1975.
27. Cowing, C., Schwartz, B. D., and Dickler, H. B. *J. Immunol.* **120**, 378-384, 1978.
28. Golstein, P., Luciani, M.-F., Wagner, H., and Röllinghoff, M. *J. Immunol.* **121**, 2533-2538, 1978.